Published by *Access Press*
P.O. Box 132, Northbridge, Western Australia.

Typesetting, layout and design by *Typeset Graphics.*

Printed by *Frank Daniels Pty Ltd,* Western Australia.

National Library of Australia
Cataloguing in Publication data
Brodie
ISBN 0 949795 89 5

Distributed in Australia and Overseas by the Publisher.

"BRODIE"

"BRODIE"

Memoirs of
Sir Laurence Brodie-Hall

This is an
ACCESS PRESS
Publication

CONTENTS

TROPIC OF CAPRICORN
WESTERN AUSTRALIA
Telfer
Warburton
MAP AREA
PERTH
Kalgoorlie
CARNARVON
PEAK HILL
BILYUIN POOL
Lake Nabberu
WILUNA
MEEKATHARRA
NANNINE
REEDY
YEELIRRIE
CUE
MT MAGNET
LAVERTON
YALGOO
YOWERGABBIE
LEONORA
Ardingley
MULLEWA
KOOLANOOKA HILLS
GERALDTON
WALKAWAY
MINGENEW
MT JACKSON
WUBIN
KALGOORLIE
DALWALLINU
BEACON
COOLGARDIE
KAMBALDA
BULLFINCH
SPARGOVILLE
Lake Lefroy
GOOMALLING
SOUTHERN CROSS
MERREDIN
Lake Cowan
NORTHAM
MIDLAND JN
NORSEMAN
PERTH
Lake Dundas
PINJARRA
HARVEY
RAVENSTHORPE
BUNBURY
Kundip
ESPERANCE
KARRI VALLEY
ALBANY
Roads
Highways
Railways
SCALE
0 100 200 300 400 Km.
0 50 100 150 200 Miles

FOREWORD

It is a privilege to have been invited to write the foreword to the memoirs of Sir Laurence Brodie-Hall, AO, CMG.

Brodie, as he is called by all who know him, was the General Superintendent of Great Western Consolidated NL at Bullfinch, Western Australia, when I joined the company in 1956. I was fortunate in having him as my first boss in the Australian mining industry. Working for Brodie was exciting because he was full of new ideas which needed to be followed up. He was years ahead of the management thinking of the time, in taking a very personal interest in the people he was responsible for, and in his ability to bring out the best in them. His leadership melded into a happy and thriving community, an isolated group of people, living in harsh conditions in today's terms, and completely dependent on a financially struggling mine. I, for one, learnt much from him.

Working for, and with, Brodie in various capacities over the following twenty-nine years after I first met him in 1956 has been one of the most memorable experiences of my time in the industry.

For some decades Brodie was the personification of the mining industry in Western Australia, particularly the gold mining industry. He can claim credit for seeing gold mining in WA virtually single-handedly through the dark times in the late 1960s and early 1970s, until the freeing up of the price of gold led in a new era of prosperity for this industry.

The story of his life is as fascinating as it is inspirational. There cannot be many people anywhere in the world who have faced and overcome so many hurdles to end up at the highest levels in their profession, so obviously enjoying every minute of it, and so clearly giving back to the community as much or more than they received from it. He has told the story in a simple and straightforward way which is typical of the Brodie many of us know so well.

Missing from his story are, however, the many recognitions and high distinctions he has so rightly received over the years. He was President of the West Australian Chamber of Mines from 1970 until 1975 and a member of the Australian Mining Industry Council for some time. For several years he was on the council of the Australasian Institute of Mining and Metallurgy, was President in 1973, received the Institute medal in 1976 and became an Honorary Life Member in 1988. He served on many Government bodies including the West Australian Environmental Protection council and was Chairman of the WA State Committee of CSIRO from 1972 to 1982. He received the WA Citizen of the Year Award in 1974, was made a Companion of the Order of St Michael and St George (CMG) in 1976, created a Knight Bachelor in 1982 and was made a Freeman of the City of Kalgoorlie-Boulder in 1989. Sir Laurence was made an Officer of the Order of Australia (AO) in the Australia Day Honours in 1993 in recognition of his significant contribution to the mining industry in Australia.

For many years Sir Laurence maintained a keen interest in the WA School of Mines and was Chairman of the school's Board of Management. He devoted himself to enhancing the status of the school and raised a very considerable amount of money to provide for its further development. He served on the council of the Western Australian Institute of Technology (WAIT) which later became Curtin University and received an Honorary Doctorate of Technology from WAIT in 1978. He was appointed a Fellow of Curtin University in July 1987 for his many years of service to education in WA.

His name is commemorated in the L.C. Brodie-Hall Administration Centre at Agricola College, Kalgoorlie, the Brodie-Hall Research and Consultancy Centre at Curtin University, Brodie-Hall Drive at Technology Park near Perth and the annual CSIRO Brodie-Hall lecture.

Recording his life story in this highly readable manner is another important service Brodie has rendered to the mineral industry and the Australian community.

I am delighted to commend this book to readers.

Sir Arvi Parbo AC
28th March, 1994.

INTRODUCTION

This book has been put together in what I imagine to be a rather irregular manner. The first piece, written in 1987, now appears about one quarter way through the text. Further bits and pieces followed at random intervals, eventually to be sorted out and put into the closest order I could get to a chronological sequence.

I had grave doubts of my ability to express myself in a narrative style which would appeal to the general public, as for several decades, apart from personal correspondence, I had written nothing other than engineering and financial reports or public addresses.

For some time people had been urging me to write my story, putting forward as a reason my long and varied life in a rapidly changing world and that people like to read about what conditions were like in Grandad's day. I had to put my writing style to the test and selecting one of the most interesting and unusual events in which I had been involved as a young man, I wrote it up. That was nearly seven years ago shortly after the October 1987 share market crash. Perhaps at that time the stimulus to write was my recognition that my fortune had been very severely reduced by that event and that I might need a supplementary income! I invited a reporter friend to read it and he assured me that my style was OK. I still didn't do much about getting on with the job other than to tell those people who kept enquiring what progress I had made, that I had made a start. Occasionally I would harvest another interesting event from my memory. Fortunately I have always enjoyed verbally recounting my early experiences and they have remained fresh in my memory which is just as well as never at any time have I kept a diary. They are truly "memoirs".

I still did not make much progress. It was my wife Jean who really got me going again. I had been writing at a very elegant antique desk which must have been made to the specifications of a person not more than four feet nine inches in height. I couldn't get my knees under it properly and it was most uncomfortable and I used this as an excuse for my slow progress.

Jean quickly demolished that line of defence by buying me the most handsome desk imaginable. Just sitting in front of it sent me on an ego trip which made writing a book appear to be a mere trifle. I gathered up all of the stuff I had scribbled at random, arranged it in chronological sequence, and got down to work seriously filling in the gaps.

Even then I had difficulty in persuading myself that anyone could be interested in reading the stories I was committing to paper until I started to read parts of it to selected visitors and it immediately came alive. In the process I came to realise how much my life confirms the truth of the old saying "that behind every successful man is a woman", except that in my case there were three women, as I must count in my mother's very strong influence. Besides inculcating in me very high standards of ethical and moral behaviour, she saw to it that I had a solid founding in the Anglican version of Christianity and taught me a strong respect for discipline. This may sound as if she was a real martinet so I hasten to add that I enjoyed a full measure of love which tempered her firm resolve that I should grow into manhood with a proper set of values and correct social manners. Her influence carried me through the difficult years when I was little more than a boy twelve thousand miles from home and exposed to a wide range of lures and temptations, most of which I seem to have resisted.

The council school I attended in England was only a few hundred yards from our home and after the world war when mother was no longer doing her bit for King and Country working in a munitions factory, she would meet me at the school gates at the end of the day and take me for quite long walks during which she would discuss my lessons, supplementing what I had been taught from her own store of knowledge which was quite considerable as she was an avid reader whenever she had time.

Politically mother had leaned towards socialism. I guess that her continuing struggle to raise a family on the meagre wages of a police constable and to provide for their further education had accentuated her feelings of injustice towards a political system which condoned a great disparity between the "haves" and the "have nots" and did little to change it. She was a very articulate person and at one time was under pressure from a family friend, who was something in the Labour Party, to nominate as a candidate for the local constituency, a proposal she declined. Mother was very much on the side of the underdog, an attitude which has rubbed off on me and left me vulnerable to the appeal of anyone in need of help.

Looking back, one of mother's main characteristics was her invincible optimism from which I inherited the philosophy that "everything turns out for the best", a belief that the major events in my life tend to confirm. She had a great store of encouraging adages which she would quote with great conviction when suitable opportunities arose. I find that seldom a week goes past without one of these flashing through my mind. In particular, "never say die until you're dead" has inspired me on many occasions to hang on until things have changed for the better as they nearly always have done.

Although my poor performance at secondary school was a big disappointment to her, she was very proud of my later achievements. She herself would have made a great success of life in Australia had things gone according to plan but when Dad died and she established a line of spiritual communication to him through a medium who lived in close proximity to her, nothing could persuade her to sever that connection.

The second woman was Dorothy Jolly whom I married in April 1941 and, without benefit of honeymoon, took her straight back to Kundip where I was managing a small mine. She had every reason to consider the first part of our life together totally unsatisfactory and maybe she did. Had she complained I would certainly have agreed that she had good cause. We were at Kundip together for eight months during which time I was so engrossed in the impossible task of trying to make the treatment plant perform to the design specifications that it became a seven days a week task with most working days being at least twelve hours and very often much longer.

It was a very small community and although the few women in it were very nice and good-hearted their lack of education made them inadequate company for Dorothy. The lack of sporting facilities was also a bind as she had been accustomed to playing both golf and tennis since she was a young girl. Being newly married there were no children to occupy her time and during those early months she led a very dull life. Dorothy was aware of my intention of joining the military forces if and when I was accepted as we had discussed this before we were engaged and although she opposed warfare in principle, as the war progressed and the allies continued to lose ground she came round to share my views and agreed that it was the proper course to follow.

By the time I was free to enlist Dorothy was expecting our first child. On leaving Kundip we stayed briefly with her parents who were by then residing in Nedlands. While waiting for the big event we looked around for a house for Dorothy and found a small furnished cottage at Gooseberry Hill at a rental of £1 a week. Our daughter Lesley was born on 21st January and when Dorothy came out of hospital I installed her in the cottage, offered myself to the recruiting officer and was enlisted in the AIF on 2nd

February 1942. From then on until March 1948 when the family, increased by the addition of two boys, Robert and Phillip, rejoined me at Norseman, she was "head of the family", in fact, if not in common usage of the title.

Her life progressively became more social, she developed quite a circle of friends at Gooseberry Hill, but it was not until she brought the family to Norseman that we had any real family life and I got to know my children. My transfer to Melbourne office gave her a much appreciated change of scene.

The seven years at Bullfinch should have brought our married life into a more normal routine but events directed otherwise as I describe later and the challenge inherent in the rapid growth of Western Mining Corporation after we moved to Kalgoorlie maintained the pressure. I recall with shame and humiliation that during a period when WMC's efforts to get into the iron ore export business were at an important, but not critical, stage and I had just taken the family to Perth for the annual three weeks holiday, I couldn't stand the inertia and sent myself a telegram informing me that I was urgently needed back on the job whereupon I left them to take their holiday without me. This was the tenor of our life together which conformed to the general pattern we had settled on during an interesting discussion between us shortly after we became engaged. Apparently it had been her father's custom, no doubt heartily encouraged by her mother, to get up in the morning ahead of his wife, light the fire – wood stoves were the norm in those days – and take her a cup of tea in bed. This was quite opposite to my views on the way our household would be run. I told her that we would start each day together. I would do my utmost to create a lifestyle for her which would justify her selection of me for husband and in the normal course of events she would attend to all domestic chores and do everything necessary to help me attain my objective. I think my actual words were "your job will be to look after me, mine will be to bring home the bacon". We both lived up to the spirit of that initial contract. In honouring that understanding I never took a major decision that affected our private lives without first gaining Dorothy's agreement to it. It was to our mutual satisfaction that when we moved to Perth in 1964 we purchased a handsome old house at Gooseberry Hill which had been built in 1910 by Fred Dempster, a pastoralist in the Carnarvon region. It was built in the colonial style, high ceilings and spacious rooms which Dorothy had a ball redecorating and furnishing. She had a flair with such things and we were both very pleased with the result which provided an appropriate backdrop against which she presided as a gracious hostess on the many occasions which my role in WMC and other organisations generated. I tried to publicly acknowledge the extent to which I had been dependent on her when I made my acceptance address on receiving the medal of the Australasian Institute of Mining and Metallurgy in 1976. Included in some advice I passed onto younger members of the Institute were these remarks:

"When marrying, choose a girl who was born in the country, for preference in a mining town and who other than having lived in the city whilst picking up a couple of degrees has grown up in the country and just adores the thought of spending her life in similar surroundings. Whilst on the subject of her education, a degree in domestic science is a valuable asset as you will need lots of stoking with the right sort of food, well-cooked and tastefully presented while you are manfully battling your way to the top. Before accompanying her to the altar confirm that she is willing to be solely responsible for managing the household, the upbringing of your children and that she will not make any demands upon you when you wish to work back at the office all night or over the weekend or when you are at home resting from those labours.

"Make sure that she is an efficient and charming hostess, a tireless worker for the community and sympathetic confidante. When the pressures of the job become intolerable you will need a soft shoulder to cry on and someone to bolster your ego. With the passage of years, during which your wife has shielded you from all domestic problems, confirmed your modest belief that you are an unselfish genius combining all the leadership skills of Winston Churchill, the financial wizardry of a Rothschild with the personal charm of Clark Gable and through her social graces has assisted you to become a big shot in the organisation, you will be ready to take on some big time community service."

Dorothy proved to be all of those things. She was very gregarious and developed a lively social background to our lives wherever we lived. I felt that I had fulfilled the promise to do my best "to bring home the bacon" when in 1971, on Christmas Day, I gave her a top-of-the-range sports model BMW. There weren't many such cars around twenty-three years ago and she was thrilled with it. It suited her personality admirably.

I first met Jean Verschuer, the third woman in my life, in 1946 when as a single woman she celebrated her 21st birthday at her parents' home in Gooseberry Hill. Jean went on to marry and raise a young family and when they became old enough to take themselves to school and back she started her professional career as Jean Verschuer Landscape Architect. She quickly established a very good reputation in that field which was in its infancy in Australia. Jean was a founding member of the Institute of Landscape Architects in 1965, and when WMC built new metropolitan premises at Belmont in 1966 she was engaged to design the grounds. The development of Kambalda was not far off at that time and Jean, whose forté was in the development of arid area sites, was the logical person to design the new town. Ably abetted by John Oliver, the Resident Manager and Lindesay Clark, the Chairman and Managing Director, both of whom were keen to see the development of the town in a bush setting, Jean was able to create a new concept for the landscape design of remote mining towns which was adopted by other companies when the development of their operations reached that phase. Alcoa of Australia, in which WMC then had a one third equity, utilised her services at their new town of North Pinjarra. In these circumstances Jean worked closely with me and when early in the 1970s at the time of a serious economic downturn in Australia she was offered the position of Landscape Architect at the University of Western Australia I strongly recommended that she take it as the funding there would be very secure whereas in the mining industry when profitability declines, the peripheral services are the first to be reduced. She wisely took my advice but the changed employment circumstances did not affect our association as I needed her advice on the development of my garden, upon which I was spending a lot of time and money, and as she lived close to us she and her family became frequent visitors to my house.

It was a dramatic coincidence that her husband Barney and my wife Dorothy died within five months of each other. I shared Jean's grief when Barney died in September 1977 suddenly of a heart attack and in turn gained support from Jean when Dorothy was undergoing the terrible metamorphosis which overtakes people when they are slowly dying in great pain.

I depend for emotional stability on an intimate association with a person of the opposite sex. My friendship with Jean, which went back over thirty years, offered a good foundation for a second marriage. We married in 1978. Our marriage has been a great success. Our respective children have integrated well into the extended family. In the new circumstances I no longer wished to remain in Gooseberry Hill and as Jean found the daily trip to the University onerous we sold our houses and moved to the city where we were fortunate in buying a house adjacent to Kings Park overlooking the Swan River.

Jean resigned from the university in 1980 very much in response to my request as I thought that we should both begin to lead a more leisurely life. That didn't eventuate. We both seemed to change paid employment for numerous honorary tasks. When in 1982 I retired from the various mining company boards I served on and ceased to have a secretary, Jean stood in. She had learned typing for a term at Presbyterian Ladies College when she was seventeen. Typing must be similar in some respects to bicycle riding, once learned never forgotten, for with the aid of a computer and word processor software she rapidly became a very efficient typist-secretary. She runs the business side of my life with great enthusiasm and efficiency and leaves her fingerprints on practically everything in which I become associated, as secretary, chauffeur, hostess, companion, making sure that my life flows with the maximum of comfort and the minimum unnecessary incursions into my privacy. She is a wife without peer.

There have been other people who have helped me along the various periods during my career whose assistance I have acknowledged in the story. They include those who converted my handwritten manuscript through successive drafts into the final proof. Ms Penny Draper, who became a friend when she arrived in Perth on exchange from Sheffield University to the University of Western Australia, typed the first one third during her term. Margaret Hooker, who had been my secretary for eighteen years with WMC, stepped into the breach for the next one third, and a young friend, Maria Pascoe, who I met when she was employed by Scitech Discovery Centre, took on the tedious task of finishing the final third and correcting the several drafts. Mrs Coral Wager, a close friend over many years, responded to my request for a map relevant to the areas to which my story refers. To these friends I give my grateful thanks. I am pleased that it is now finished but having become accustomed to writing I wonder what I will find to substitute for it.

Gilbert Ralph, who had been my technical assistant for several years before being transferred to WMC's head office in Melbourne, had also given very valuable assistance. In his present position as WMC's Manager Corporate Publications, company historian and archivist, he was able to correct my manuscript for historical errors and supply me with appropriate photographs from the archives. In addition, he found in the final draft, enough typographical errors and some omissions to make me aware that as a proofreader I need more practice. I acknowledge that help with grateful thanks.

To the key people whose friendship has been special in my life, Frank Espie Snr, Lindesay Clark, Arvi Parbo, Roy Woodall, Odwyn Jones, Gilbert Ralph, Neil Tuckwell and John Brophy, I acknowledge their friendship and give them my concluding thanks.

Portrait of the author.

Chapter 1

Although I doubt if it registered at the time, I must have been a big disappointment to my parents. The early promise I showed at council school, where I was usually amongst the top places, failed to materialise at secondary level and I squandered the sacrifices my parents made to keep me there. It is difficult to explain why there should have been such a falling off. I had no competing activities to distract me from my studies. I played no sports and had no hobbies and most of my leisure time was spent reading or practising the violin. It must have been either sheer laziness or adolescent lethargy, combined with an infatuation I had for outside adventure – adventure of the romantic kind as purveyed in Arthur Mee's Children's Newspaper – exploration in the depths of the Congo, the throb of native drums passing on the message of the white man's approach, the discovery of fabulous deposits of gold and precious stones. It is difficult to take study seriously when the mind is occupied with thoughts of being carried up into the jungle by native bearers, or whilst mentally enduring the rigours of a stiff bout of malaria.

It was particularly rough on my parents. The wages my father brought home as a police constable were meagre and to pay my way at Sir Walter St John's secondary school and keep me clothed to a standard that would not unduly humiliate me in the company of children of more affluent families required extreme frugality in their own lives. Mother was still wearing for best a costume she had bought when she was married, twenty years previously, and Dad was no better off.

Both my parents had been in service with one of the landed gentry and after marriage invested their savings in a small business for which, alas, they had no previous training. Regrettably they lost their savings and father was fortunate to be enlisted into the London Metropolitan police force, for which the weekly wage at the time was twenty-nine shillings. As the family grew, mother supplemented these slender earnings by taking in laundry.

I was the last of the family, born on 10th June 1910, and by the time I started to develop a permanent memory bank, mother was doing her bit for both the country and her family during the 1914-18 war by becoming a munitions worker. I think she operated a lathe and turned out shell cases, or some part of them. There were no child care centres in those days, but fortunately we lived close to the school I attended and although there was a lengthy gap of time between finishing breakfast – mother's shift started at 7.00 a.m. and she had about an hour's travel to work – and the start of school at 9.00 a.m., with the same sort of interval between the end of school and mother's return home in the evening, I do not recall any sense of personal hardship in this regimen.

On looking back I think that father's work must have been rostered to include some evening shifts as I have strong recollections of the German air raids. If these were in the evening we would sometimes take refuge in the Latchmere Road Police Station which was close to our home and heavily sandbagged. On one occasion, apparently when Dad was on duty and mother at work, we saw from our front room windows, picked out by searchlights, a German Zeppelin attacked and brought down in flames, breaking in half as it fell. It must have been fairly high as it appeared as though it would fall not far from us when in fact it fell on the other side of London. Our flat was on the third floor (the top) of the tenement block and we enjoyed a grandstand view of the event.

It was an austere time in which to grow up. I was four years old when the war started and food was strictly rationed. This, combined with a strong instinct on mother's part to save for a rainy day, meant that we lived on a spare and very simple diet. A large plate of coarse oatmeal porridge was our invariable breakfast and the small amount of meat we had was very largely what is now disparagingly referred to as offal. Mother knew all about the value of vegetables and in so far as they were available, we had our fair share. Bread with dripping or Australian jam, Jones IXL quince jam features strongly in my memory. I find it interesting that most of the food I had as a child still stands high in my regard now, when alternative fare of what is deemed to be better quality is readily available. As a result of a relatively fat-free and certainly sugar-free diet, I grew up with a skinny, strong frame and a good set of teeth.

Somewhere along the line I absorbed a good sense of the value of money and the importance of thrift. Toys were not readily available, nor in our household was there the money to spend on them, other than at Christmas, when each member of the family received one toy, the total for the year, but we grew up to be a family skilled in draughts, chess and a variety of card games, as well as other forms of self-entertainment.

Of our family of four children, Douglas, the eldest, was eight years older than I, Irene, called Rene, had six and a half years' seniority, and Frank, three years. We got on very well together considering what kids are like. Even when I was young, from five to six years old and onwards, we spent much of our school holidays as a group either in museums or one or another of the half a dozen parks or commons within reasonable walking distance, say four to six miles from home. On special occasions when setting out for the day, each with a packet of sandwiches, one of bread and dripping and another of bread and golden syrup, we would each be given a halfpenny. Douglas was the self-appointed group treasurer. It was our rule that a ha'penny was reserved to buy mother a chocolate and on those occasions when we went further out such as Wimbledon Common or Richmond Park, a ha'penny would be reserved for Rene for a tram fare from the top of East Hill to home. Rene being a school girl qualified for a half fare but the real purpose was to take me on her lap. Although I was rather mature to be classed as a babe in arms and had no right to a free ride we generally got away with it. On rare occasions a ticket inspector would board the tram, the signal for us to alight at the next stop.

We always talked about Wimbledon Common being four miles from home and it may well have been so. I know that after playing all day on the common or picking bluebells, buttercups or blackberries, which grew in profusion there in those days, the return trip would have been impossible for me without the ride home. The remaining penny would be spent on those sweets which could be bought either in numerical quantities divisible by four, or in units costing a farthing each. Aniseed balls, a lolly which proved very resistant to erosion even from vigorous sucking, could be purchased twenty-four for a penny.

During the post-war years Douglas and Rene became teenagers and grew out of the close family association and Frank and I formed our own individual friendships. The museums at South Kensington, Science, Natural History and the Victoria and Albert, became a great attraction for me and I undoubtedly added substantially to my store of knowledge from those visits.

Following the end of the war specialty items of food came back onto the market and by Christmas 1919 had reduced in price to the level at which we could afford some. I came home that Christmas Eve to find what appeared to be a huge mound of fruit on the kitchen table. On reaching out my hand to tentatively touch them I heard my mother's admonition, "You may walk all

Grandpa Brodie

round but you may not touch", an expression which subsequently became a well-used phrase in the house whenever similar circumstances occurred.

Subsequent Christmases were spent at Grandpa Brodie's house in York Road. Gramp was a successful master builder and it was a lasting disappointment to him that he had no son and heir to carry on the business and the name. He had seven daughters, of whom mother was the second youngest. His wife died at a comparatively young age of cancer I believe, but it may well have been due to giving birth to a succession of children in a minimum of time. Anyway Gramp was very keen to have the Brodie name carried on and it was incorporated in the name of the male children of my mother's family.

I know very little of Dad's side of the family. Apparently Grandfather Hall had died when quite young but not before he also had fathered a large family of girls whom Grandmother Hall had brought up. She had a little shop in Woodford, Essex, from which she sold ribbons and silk thread and such like. I don't think she had many customers as most of the time the shop was unattended, Grandmother being alerted to a customer's presence by a handbell secured to the front door by a piece of springy hoop iron which caused the bell to jangle noisily whenever the front door was opened. I found her a frightening old lady and was not at all sorry that she lived on the other side of London to us as such a long trip was only taken on rare occasions and we didn't see much of her.

The trip across London on Christmas Day from the Battersea Town Hall on Lavender Hill to Kings Cross by number 77 bus, maintained the level of excitement which had been building up for the previous week. London buses in those days were open to the weather on the top deck where we kids always preferred to ride. To protect the top deck passengers from the rain a canvas apron was provided which fastened to the back of each seat and which could be drawn up in front of the passenger in the seat behind. It kept the rain and snow out whilst in motion but unless great care was exercised would decant the accumulated load of snow or water on the passenger's legs when alighting. The buses ran on solid tyres and bore the sign "not to exceed 12 miles an hour". I doubt if they ever reached that speed as it was usually quite easy to catch them by running them down providing they hadn't got too large a start.

Most of the high roads and those in the city were paved with jarrah blocks, sealed over with bitumen. Some time after the war when road building methods had improved these were pulled up and I would be up there among crowds of other people, filling a sack with as many as I could lug home for firewood.

The Christmases spent with Gramp after the war were the highlight of the year. The lead-up to Christmas was always exciting. The commercial farce of Christmas was not evident until much closer to the actual day than has become the custom nowadays and because of the shorter lead-up Christmas was more intense. We children were allowed to draw a small amount of money out of our savings bank account to spend on presents usually either in Marks and Spencers' penny bazaar or at Woolworths, where no item cost more than sixpence. We each bought a small present for the other members of the family. I guess the five presents that each bought would have cost in total not more than two shillings. Mother and father bought one present between them for each of us. Christmas dinner was roast chicken, in those days a rare treat, and included, of course, the traditional plum pudding which was made and given the initial cooking several months before Christmas Day, one of several prepared in advance to be eaten as a special treat on the family birthdays.

Christmas was the one occasion upon which we were allowed to stay up until we fell asleep and were carried to bed. Normally throughout the year I was sent to bed by seven o'clock on the grounds that I was a sensitive child and highly strung and needed all of the sleep I could get! I endured this regimen by promising myself that when I was grown up and had left home I would stay up until midnight every night although, when I started work it was a different story.

I don't quite know how it developed but by the time I was twelve I had become an ardent church goer. Not to the church our family attended, and in which I had been confirmed, but a small church in the same parish where I had attended Sunday school since I was very young. From twelve years onwards until I left home my regular Sunday routine was 7.00 a.m. early communion, 11.00 a.m. morning service, 3.00 p.m. Sunday school, 7.00 p.m. evening service. The congregation were a nice friendly lot and I suspect the social component attracted me as much as the religious services.

About this time I joined the Boy Scouts. I enjoyed the regular weekly evening meetings held in the loft of a disused building when we seemed to spend most of our time drinking cocoa and learning to tie intricate knots which I am sure very few of the troop had a need to use afterwards although I found some which were useful in my later life, in the bush, when I went to Australia. Most importantly we were taught strict codes of ethics and morals which I believe stuck and reinforced what I had learned, by example, at home.

Being the youngest of the family, I guess I enjoyed privileges that a more severe shortage of money had denied my two brothers and my sister. When I was about ten years old, I started to learn the violin. I really wanted to learn to play the piano, but a violin had been bought for Frank, the brother next above me, and when he decided to transfer his attentions to a tin whistle, I had to take on his abandoned fiddle. This is an early example of my belief that everything turns out for the best, as subsequently I toted that violin to many distant places around Australia where it won me friends and earned me money when the economy was depressed and work was hard to get. Comparable benefits would not have existed had I learned the piano. During the Depression, in the absence of wage-earning work, the violin kept me financially afloat and also opened the door to the homes of many people who helped to advance my career.

At about two years old, my left eye had turned inwards in a very pronounced squint, which glasses failed to correct. This was unfortunate, as it led to an event that put me under a great deal of stress for many years. When I was about eight years old, it was decided that the cross-eye should be corrected surgically. Those were days long before the welfare state but many of the great

My mother.
On the back she has inscribed a wish that the Lord bless and keep me always.

My father.
The inscription on the back of his photo is more romantic.
He advises "hitch your waggon to a star".

houses had a Almoner on their staff to disburse money for such things to those in need. If I remember correctly, the operation, which was attempted by one of England's leading eye specialists at the Moorfields Eye Hospital, was paid for by Lord Derby. I say "attempted" because I put on such a performance that the surgeon gave up soon after he started. I can recall the scene very well. The operation was to be performed with a local anaesthetic, while he described to tier upon tier of medical students surrounding the operating table how he would proceed. He didn't get very far – when he got to the stage of announcing that he would "now make an incision in the conjunctiva", I let fly and was immediately dismissed.

As a result of the psychological shock, I developed a speech impediment which stayed with me until my middle teens. It wasn't a good old-fashioned machine gun repetition of difficult consonants, but a horrible, soundless grimacing, as I tried desperately to articulate words that would not come out. There was an exception to this condition. I had some talent in recitation and while I could only converse with extreme difficulty, I could recite at length without any hesitation. Most of the school teachers could not understand why this was so and tried, unsuccessfully, to knock some response out of me. I think it was a matter of confidence – I knew that my recitation was good and the satisfaction I derived from being taken to some of the senior forms to recite Shakespeare or Keats or whoever, inspired the necessary confidence.

Perhaps the speech impediment was part of the reason that I flopped at secondary school, as the teachers gave up trying to get answers to their questions from me, so I lapsed into day-dreaming of exploring new continents. When I left home and was thrown onto my own resources, it took three years before I overcame the speech problem, during which time I put myself forward in public to sing, recite, play the violin and finally, to make short extempore speeches. I had many humiliating experiences in the process, but by then I was hardened to ridicule. When I was at council school I had cut a sorry figure with an eye in the corner, steel rimmed glasses, heavily swathed in cotton wool over the nose and the ear lugs to stop the frames from cutting into my skin, and the terrible speech defect. I was an obvious butt for ridicule. Strangely even at this late stage in life, after much experience, when I get to my feet to make a speech there is a cell in my brain which generates a small doubt that I might get stuck on a word and find myself grimacing soundlessly at my audience. Thankfully, it has never happened.

Mother and father were a devoted couple, a condition which

adversity seemed not to shake. They were both physically big. Mother was five feet nine inches and Dad six feet three, but mother was the dominant party. Dad was mild-mannered and gentle, whereas mother was forceful and authoritarian, meting out quite severe corporal punishment whenever she considered it called for. Anything of a minor nature was dealt with by a sound box on the ear, for major offences the preferred instrument was a very flexible cane, purchased from what we called the "Oil Shop", a cramped little emporium that sold mostly hardware and china, but also bulk jams and syrup and pickles, which could be bought by weight if the purchaser supplied the jar. A penny's worth would be the usual amount served. It was a great humiliation to have to ask for a penny cane and we usually covered up our embarrassment by citing in a loud voice a fictitious purpose for which the cane was required. In these circumstances I grew up with a very healthy respect for discipline and authority.

As the youngest of the family, I fell heir to most of the domestic jobs like shopping and cleaning the knives and forks. Those were the days before stainless steel and the cutlery had to be cleaned weekly, the knives on a strip of linoleum, about two feet long fastened to a board, using "Wellington's Oaky Polish". The forks and spoons were cleaned with fine chalk. For this task I received sixpence weekly which I was made to save. I also did the daily shopping for a bachelor taxi driver who lived nearby and unfailingly rewarded me with threepence. His daily diet consisted of one of three choices, each purchased in one pound lots – either a slab of rump steak, a gammon rasher of bacon, or a chump chop, this being cut from the thick end of a leg of mutton. He worked the night shift and this was his "breakfast", eaten at about 6.00 p.m. and washed down with a quart bottle of Guinness stout. When I left home I had accumulated from these, and similar services, a little under £10.

When it became apparent that I was unlikely to establish a satisfactory record at school, mother, being a practical person, laid the foundations for the prosperous and rewarding life that she had mapped out for me by invoking plan two. I should go to Australia and learn to be a farmer and when father retired three years later on a pension, after twenty-five years in the constabulary, I would take up a selection and the family would all join me in Australia to develop it, using mother's accumulated savings as a capital base and Dad's pension for working expenses. My brother Frank, who had already been in South Australia for three years, would join us. It wasn't quite the life in the African jungle that I had day-dreamed about, but it offered a similar element of excitement and adventure and I was pleased to renounce the life of a London schoolboy and start a life in Australia.

The fateful day dawned when I boarded an Aberdeen White Star Liner, taking with me a massive cabin trunk, which would have admirably served a film star going on location. This was crammed full of the most inappropriate clothing, as recommended by the staff of Australia House. It was a great waste of my parents' money that they could ill afford and which if I had worn after my arrival, would have made me a laughing stock. In the 1920s the Commonwealth Government had an immigration scheme going, under which selected youths were admitted to Australia and given a ten week orientation and training period on a teaching farm out of Sydney, before being found a job on a farm and turned loose in the Australian community. The total cost to the parents, other than that associated with the wardrobe of useless clothes, was £10.

Before the successful eye operation.

To the best of their financial ability, mother, father and even my eldest brother Douglas, who was working with Unilever as an industrial chemist, showed me the sights of London in the months preceding my departure. Until that time my knowledge of London had been confined to the museums, where I spent a great part of my school holidays. During those last few months I was introduced to the great art galleries, to St Paul's and Westminster Cathedrals, the Albert Hall, Covent Garden and Drury Lane, the Houses of Parliament, the Changing of the Guard and many other institutions and events that are every Londoner's heritage.

Most importantly, I had another visit to Moorfields Eye Hospital. Dad accompanied me to the hospital on the top of a double decker bus, gently telling me that "this time you must be a Briton and grin and bear it". The operation was successful, but as I am very long-sighted, I have had to wear quite strong lenses ever since, which later proved to be a considerable handicap when I first tried to enlist in the AIF in 1939.

After the successful eye operation.

The dockside farewells were accompanied by promises to write home regularly, which I regret to say I honoured more in the breach than in the observance and my parents were left waving forlornly on the dockside at Tilbury, while my boat disappeared over the horizon. It was an uneventful trip. I don't know what the shipping line charged the Australian Government for the passage of the group of thirty or so young chaps travelling under the scheme, but whatever it was, I am sure that it was too much. Between first class travel and steerage there is a great difference.

Scheyville, as the Training Farm was called, was close to the town of Windsor in New South Wales and turned out to be a pleasant experience. The only permanent memory I have of my stay there was a strike staged by our newly arrived group, on the grounds that the food was not fit to eat. I refused to join the strikers. I had no complaint with the food and although the fare at home had been very plain and sometimes sparse, I am sure that it was a lot better than that to which most of the others were accustomed. I endured being called a scab and other standard union epithets and for my pains was thrown into a very muddy dam whilst wearing my good clothes.

I refer to the Scheyville period as "when I took a crash course in Agricultural Science". It was of ten weeks' duration, with each week devoted to becoming acquainted with some aspect of farming. The week with horses culminated in driving a pair of very well trained Clydesdales in a single furrow plough. The ground had been turned over very many times, there were no concealed roots to cause problems and the horses had done it all so many times before that they needed no control. By the end of the course I had acquired the skills necessary to get a job as a farm labourer. I could tend horses and drive a pair in a single furrow plough, an archaic occupation that I was never called upon at any time to perform, milk cows, feed pigs, clean out chicken sheds, herd sheep, stook wheat, pick oranges and had learned how to kill and dress a sheep and a pig, all of which qualified me for my first job at £1 a week and keep.

I found that while at Scheyville I developed the swearing habit. I guess it is, or was, characteristic of men living in the bush, but it is probable that it was rife among the party I came out from England with, of which I was the youngest. I was pulled up very sharply one day by the chaplain for using language that really was "over the fence", as I recognised when he repeated it to me and I eliminated the very rank epithets and adjectives from my vocabulary from that date, retaining for use when occasion warranted "bloody" and "bastard", as evidence that I qualified as an Australian.

Against the background of my London experience, I found the Australian bush beautiful and the open air life and physical work most satisfying. I looked forward to my first job with enthusiasm.

Chapter 2

There was no marching out ceremony at Scheyville. When a job was found for me, I was taken in to Sydney Central Station, given a third class ticket to my destination and turned loose to find out what the future had in store. Until then I had always been somebody's responsibility. Now I was completely on my own.

I surveyed the train carefully and selected an empty, non-smoking third class carriage, hoping that I would have it to myself. I almost did, but seconds before the guard gave the driver the green light, a figure of a man appeared staggering under the weight of a heavy suitcase, a bulging sugar bag from which the sound of glass on glass could be clearly heard, and too much drink carried internally. He threw open the door, dumped his gear on the floor and subsided heavily on the seat as the train lurched forward on its journey north. It wasn't long before he pulled a bottle out of the bag, whipped the crown seal off with his teeth and took a long swig, after which he wiped his shirt sleeve over the neck of the bottle and extended it in my direction, with the terse command "'Ave a drink!" Somewhat subdued by his wild appearance and aggressive manner, and in view of the fact that I had no experience with alcohol, I timidly declined. It was not a wise decision. I should have tried a little make-believe drinking, for my refusal and my accent brought down the full force of his wrath on my head. "For Christ's sake", he said, "a f . . . ing pommy, a f . . . ing wowser pommy. Here you, 'ave a drink, you pommy bastard!" It was an unpleasant trip. I quickly began to wish that Australia wasn't so big and that train journeys didn't take so long, and that someone else would join us in the carriage to take some of the pressure off me, but no-one did. I stuck it out, the abuse became more obscene as the contents of the sugar bag became depleted, until eventually he fell asleep. In the early hours of the morning the conductor put him off at his destination. As the train pulled away, I could hear him complaining about the pommy bastard. It was par for the course in those days – immigrants were not welcome.

My destination was Delungra, a small town a few miles west of Inverell. I nearly got off to a bad start. I did not have much sleep due to my concern over my fellow traveller and when the train reached Moree, where I had to change trains, I was sound asleep. Fortunately the guard was not and he woke me up and got me on to the other train a minute or so before it left. My first employer and his wife evinced the same contempt for me as a pommy as my drunken travelling companion. However, they fed me well and I had a small wooden hut, about eight feet square, under a big old pepper tree, which was comfortable enough. They were an elderly couple, with whom I failed to develop a rapport. There were no young people around during the week, which did not matter very much as I spent most of my time working. My first job was driving eight horses abreast in a six furrow mould-board Shearer plough. It was exacting work. As there had been a drought in the previous season, there was practically no feed on the ground and the farmer was too mean to buy chaff, which was scarce and consequently very expensive. To keep the team going, I had to flog them with a whip almost incessantly. It is characteristic of farmers that they know precisely how many "rounds" of the paddock should be completed each day, regardless of whether it was on the long runs, when the work was around the perimeter, or when it was on the short runs, as the work approached the centre. The first query at the evening meal was "how many rounds did you do today?" Being soft-hearted, I was sorely tempted to give the poor emaciated animals rests during the day, but the fear of being fired, when I was so new to the country, kept me steadily flogging away at them. It was very monotonous work, starting at 5.00 a.m. bringing the horses in, getting the collar and hames on them, and tending to their collar sores, finishing at about 7.00 p.m. by giving them a rub down and putting ointment on their sores, before turning them out to forage. It was made more distressing as I grew to be very fond of them. They each had a personality and they must have made quite an impression on me, as I can remember the names and characteristics of six of them now – Trimmer, Lion, Colt, Gypsy, Bluey (the furrow mare) and Blossom. "Lion", gaunt and Roman nosed, just could not be stopped from working, always out in front of the rest. It caused me great distress to watch him. Blossom either had something wrong with her waterworks or stopped intentionally at least a couple of times around the paddock to pee, just to have a rest. Bluey had her own slow, reluctant pace, which wasn't fast enough for me to be able to satisfy the boss's requirement for each day's work and consequently came in for more than her share of the whip, made from a pine sapling about three metres long, with a two metre thong.

I don't know why they sacked me. Perhaps they didn't need

anyone until fallowing started three months later. I had a reasonable amount of money by then and was not sorry to leave them. The job had lasted twenty-six weeks and my savings increased by £25/15/0, five shillings having been spent mostly on postage stamps.

It had been a humourless period with few memorable events. One of these, although tinged with sadness, was the day Colt died. Colt was one of the team, a young horse as suggested by his name. Following some sustained rain at the end of autumn, he was not to be found for a couple of days. Eventually he was located lying under a big old gum tree, his feet feebly digging at a largish hole they had already made. His eyes were bleary, his nostrils covered in mucus and his breath rattling and rasping in his chest. He was obviously far gone with pneumonia, but the farmer decided he would try to save him. A block and tackle was brought and attached to the tree, a couple of wheat bags sewn together were worked underneath him, secured to the pulley block, and after a bit of puffing and grunting, we had him off the ground and hanging hump-backed in the sling. With a bit of massage he was at last able to stand unassisted on his four legs and a start was made to walk him to a shed about a mile away. There were four of us and we had a leg each. I had a hind leg, with my shoulder strategically placed under his hip joint, to take the weight if his leg gave way. It would have been better had we been five instead of four, because every little way one of his legs would let go, requiring someone to relinquish a leg that was still carrying weight, and dash round to straighten up the leg that had given way, hoping that the other legs would stand firm. Sometimes this would happen four or five times without moving from the spot, but after several hours we got him down to the barn and slung between the propped up shafts of a dray, where he was given a bucketful of corn cobs that had been stripped of their corn! What the poor animal needed was a hot bran mash with half a bottle of whisky. It was no wonder he was found dead in the shafts in the morning.

On another occasion, when the farmer decided to bring his sheep back from Moree where they had been agisted, I drove the tucker cart, which also carried the temporary yards. I had to erect these each night and pull them down and stow them away each morning. I was also the cook. It was fortunate that after the war and until I left home, I spent a lot of time with mother, when I learned how to cook – in her style of course. I don't think she ever weighed anything, ingredients were measured out by the cup or spoon, or handful, and a recipe book was unheard of. Everything always tasted very good and for the years that I spent cooking for myself, this experience proved invaluable. It was certainly handy when I was the drovers' cook.

During the drove our flock of sheep had a head-on confrontation with a herd of cattle, in the middle of an area very heavily infested with prickly pear. After a consultation with the cattle drover, he let his cattle into the pear to allow our sheep to get through. We then went back and helped him get his cattle out of the pear and through to the other side. It took all day. Since then the Cactoblastis beetle was introduced and the pear, which covered vast areas of Queensland and northern New South Wales, has been completely eradicated.

My departure from the farm and my farewell of my employers was accomplished without any torn heart strings and I made the return trip to Sydney in less traumatic circumstances. The train arrived at Central Station at about 6.00 a.m., and not having a place to stay until I found somewhere, I "cloaked" my luggage and started to walk around the city. I shortly found myself in a side street, looking at the contents of a pawnbroker's shop window, when a pair of red-eyed, unshaven and otherwise dirty old men sidled up and begged a couple of shillings for "when the pub opens". I had no intention of parting with any of my hard-earned money for such a purpose and told them so in what I suspect was rather a quavering voice. I gave in very rapidly when one of them produced a razor from its case, opened the blade, waved it in front of my face and asked "would yer like to buy a razor". I thought I had fallen into the hands of a razor gang and lost no time in finding two shillings and beating a hasty retreat. In my fright, I failed to recognise that they were two harmless old drunks, who were waiting for the pawnshop to open to pawn the razor.

Jobs were easy to get and within a couple of days I was heading west to Trundle in the central wheatbelt of NSW, and a succession of jobs. The social atmosphere with these new employers was one of general disinterest. I had to be housed and fed and paid but, as a pommy, not very much. During that time the Jack Lang government brought out an industrial award for adult farm labourers of £2/15 a week, plus keep. Although I was only fifteen and as thin as a lathe, I was six feet three inches and passed myself off as an adult. I expect my employers realised that I was well short of adulthood, and paid me accordingly. As long as they got full value for money, and they made sure they did, they were satisfied.

One chap didn't quite conform to those circumstances. His name was Hillary Allez and he was a real wowser. I used to help a neighbouring farmer on a Sunday, a nice chap whose company

I enjoyed. Allez objected to this on the grounds that the Sabbath day was holy and for that reason exempt from work. I suspect his real reason was that he wanted me to rest up all day, so that I would be fully recharged for him on Monday. I remember that I felt very smug when he tackled me about this and I gave him some of his own back by telling him that in my book Sunday was not for being good, but for doing good (take that you old wowser).

He had a sheep dip which he rented out to some of his neighbours, an event that occurred during the short period I worked for him. It was probably around late October or November, because it got light very early and so that we could start dipping the sheep at first light, we would get up at about 3.30 a.m. for an early breakfast. The sheep were always yarded the night before and the dip filled with the initial charge of water and Cooper's sheep dip. My job was on the race. Nearly all the sheep had been through that dip before and I am sure that they knew what was in store for them, because they would prop on their front legs and the only way to get them down the race and into the dip trough was to grab them under the tail, lift the back legs off the ground and impel them forward on their front legs. Every animal was an individual contest of strength. I guess about one hundred went through each run, after which old Allez and the owners of the flock being dipped, having exhausted themselves by barking in imitation of sheep dogs to keep the sheep moving up towards the race, would sink gratefully to the ground against the yard rails to smoke a pipe and recover in readiness for the next batch. While they were recuperating, I would take the horse-drawn water cart down to the dam to replenish the water carried away in the sheep's fleeces. This was very good exercise to shake down the early breakfast, wading out a couple of paces into the dam to fill a four gallon kerosene tin with water, and then hoisting it up about six feet to pour it into the tank on the water cart, which I think held about one hundred gallons. When it was discharged into the dip to restore the level, the old fellows would clamber to their feet to begin the tiring business of barking again. The work would usually go on until midday, after which I was allowed to go to my tent and rest until it was time for me to do the milking and chop the house wood.

Harvesting started shortly after and my task changed to bag sewing. I was sewing around one hundred and fifty bags a day and intermittently throughout the day helping a carter to load about the same number of bags on his T model Ford truck. Bag sewing was usually paid on contract and had I been working under those conditions, I would have been receiving very much more than my £2/10 per week wages. Like Oliver Twist I asked for more and like

Oliver, I was denied it. "You ungrateful boy," said Allez, "you're nearly getting a man's wage already, and think of all the time off you had during the sheep dipping." My recollection was that from 4.00 a.m. until midday was a pretty good day's work and that I had done all the heavy work, while he and his pals were busy imitating sheep dogs. I exercised the good old Australian prerogative of telling the boss what he could do with his job. In the current vernacular, he was "ripping me off". There was a great deal of satisfaction in that moment and on the rare occasions when it has happened to me in reverse, I knew just what pleasure the other chap was experiencing.

It took me some time before I felt totally secure when I was camped in the bush on my own. I was frequently in this situation when doing the odd bit of clearing or fencing on piece work and not enjoying the comfort of a camp at the homestead and prepared meals.

These were occasions when, camped alone in a tent, my cooking skills should have stood me in good stead but I was still in my mid teens and to arrive back at camp after a long day's work in the field banished any thoughts I may have had of cooking anything but the easiest and quickest meal. Getting fresh food was always rather a problem. At one time for a short period I had a horse and sulky that I bought from a couple of drovers with whom I had been working. During this brief association they had not used the equipage which they sold to me on their departure, leaving me to learn the hard way that the horse, a fine looking animal which initially won my heart, had a most uncertain temperament and on irregular and unpredictable occasions he would either kick himself free of the sulky or, if being ridden, deposit his rider in the dust. Having eased his feelings or asserted his authority, whichever way he thought about it, he would then stand docilely by while the harness was mended and he was again attached to the sulky or had an occupant in the saddle. After the first occurrence, following the departure of the vendors, I realised why they had sold him to me for the knock-down price of £2. I eventually got fed up with his antics and I put him in the town pound and conveniently left him there. For a short time I drove him into the local town on Saturday mornings and bought my supplies, otherwise I depended on others to do my shopping for me. For the most part I would buy what was euphemistically called "tinned dog" and sold under the name of corned beef which, if boiled with potatoes and onions, made a tasty stew. Next on the list was self-raising flour, dripping and golden syrup, the basic ingredients for puftaloons, and then sugar, cocoa, cornflour and condensed milk. These last items being the ingredients for chocolate blanc-mange, a billy-full of which I shudder to recall would frequently constitute my evening meal. As bread went stale very quickly, puftaloons provided the carbohydrate in my diet. Easy to make – a dough made from self-raising flour and water with salt added, was cut into attractive shapes to give that artistic touch, then deep-fried in beef dripping in the frying pan and eaten with copious additions of golden syrup! What a marvellous thing a boy's digestive system is.

On one occasion I had rigged my tent over a timber frame that had been built by a previous occupier. For long term occupancy this mode is preferable to the more usual method of having external tent poles as, with the frame inside, the hurricane lamp could be suspended from the ridge pole and clothes and other items hooked onto nails driven into the side members of the frame. The bed was usually constructed by inserting two poles along the sides of a couple of chaff bags and supporting them on A frames, or at the four corners on forked sticks driven into the ground.

There not being any diversion after the evening meal it was customary to go to bed very early. On one occasion I was sitting up in bed reading Bram Stoker's "Dracula", which I suspect was tending to make my hair stand on end. It was a still night with no air stirring until a vagrant wisp of passing wind made the tent fly flap against the tent causing me to look around momentarily in nervous apprehension. I had just settled down again when a stronger puff dislodged my belt from the nail it was hanging over and it fell. When it landed it struck me with the buckle before coiling itself around my bare buttocks. Pyjamas did not figure in my wardrobe in those days and I guess I had gone to bed in my singlet. Nothing has ever simulated a snake attack more effectively. For a few seconds I was certain I had been bitten by a snake and the nervous reaction ejected me out of bed in one convulsive muscular spasm. Very humorous to look back on but not at all funny at the time.

Eventually I was taken on by a young ex-soldier, Roy Provan, to clear a paddock of scrub pine. He was a single man and had a block of four square miles (two thousand five hundred and sixty acres), some twenty miles out of Trundle. I imagine it was a war service block which he had fenced and partly cleared and stocked with a flock of breeding ewes. He was a jovial fellow and quite a character. He had represented Australia in swimming in the Olympic Games in the early twenties and had toured Queensland and New South Wales with Sharman's Boxing Troupe, following the country show circuit. Anyone who could stay three rounds with him earned £5. He had a single room cabin, with an open fire at one end on which we cooked, a table in the middle and two beds at the other end. Not a great deal of comfort, but the companionship was great. I stayed with him for a month. We shared the domestic chores, except the dishwashing after the evening meal. We played cards for several hours each night to determine who would do the dishes. I did them every night. He always ended the game while he was in front! I only stayed for a month because his prospects of building up a successful farm were shattered when a very severe bush fire, extending over a wide front, bore down on the district. Being full of initiative, he decided that he would call in the local fire fighting team and have himself appointed captain. The district farmers were happy to agree and under his direction, assisted him to successfully burn back on a number of fronts, which resulted in his property being the only one not completely burned out. It was a long day of fire fighting and when we returned to the one room shack just before dawn, he told me what a good day's work it had been and that if he hadn't organised it, his farm would have also burned. In fact he had miscalculated. In those years the rabbits were in plague proportions. When the fire went through, they went to ground. After the fire had passed, they concentrated on the only grass remaining in the district – his. Within twenty-four hours his pasture was a seething, jumping mass of rabbits. Within a week the place was eaten bare. That finished him off and my job went with it too.

I was sorry to leave him. It was a congenial job clearing scrub pine, which was the main vegetation on his property. He lived about twenty miles out of Trundle, where his fiancée lived and whenever he went to visit her, driving his old T model Ford, he would take me with him. I would go to the cinema and I sometimes had a long wait for him to finish his courting, but he was not strict about early rising, as I found out on my first morning, when following the routine I had been used to with old Allez, I called him for his breakfast, which I had on the table at 6.00 a.m. I was told very forcefully, "Go back to bed, you silly bastard". I was very appreciative too on one Sunday morning, when we had gone to town and I was sipping a weak shandy in the bar of the hotel. A big hulking lout standing next to me decided he would have a bit of fun at my expense and started roughing me up. Provan was on the other side of the island bar and when he thought this had gone on long enough, walked round, reached over the bully boy's shoulder, grabbed his shirt and as he swung him round, gave him a short right hook to the jaw that dropped him in a heap. "You had better go over to the milk bar before he comes to", he said. I wasted no time in taking his advice.

After a year or eighteen months, I heard from my mother that my brother Frank had gone to Western Australia and I decided to go over too. By that time I had saved quite a bit of money, more than enough to take me west and keep me until I found employment. But in the event, when I reached Adelaide, after uncharacteristically splashing my money around in Sydney and Melbourne, it had dwindled to the extent that, after allowing for the fare to Perth, I was so short of money that I thought it better to return to where I knew I could find a job. I must have been very low in funds, because when I found myself stranded in the railway yards of a town called Harden – it was a Sunday and no trains were running to where I wanted to go – I had no money to buy food. As a sixteen year old boy I had quite an appetite and by the time I reached my return destination on Monday night, I was really hungry. It was twelve months before I made another attempt.

I was gaining in strength and assurance all the time and becoming more useful as a farm labourer. I have a great secret pride in earning the male basic wage when I was sixteen, driving twelve horses tandem in a fourteen disc "Sundercut" plough. That was really hard work. Starting at 5.00 a.m., feeding the horses and putting collars and hames on them and after breakfast taking the team out to the paddock and having them hooked up and ready to go by sun up.

The team was made up of eleven mares and one gelding called Sammy. One morning after hitching them to the plough and climbing on board and crying "giddup", the eleven mares moved, but Sammy refused to budge. Sammy was in the front row and nothing could move him. I got between the two rows of horses and belaboured him with the spreader used to keep the trace chains uniformly apart, an idiotic thing to do, as the team could have started with me in the middle. In the end, after about one and a half hours of fruitless trying, I went to the front of the team, grabbed the offender's bridle and gave him a full-blooded kick in the belly. The result was instantaneous. He reared up on his hind legs and came at me with his ears back, front legs flailing and lips rolled back, exposing his long, yellow teeth. Blind with rage, I grabbed him either side of the bit when his front feet hit the ground and screaming "bite me you bastard, would you", I sank my teeth into his nose. It had two effects – it eased my feelings and it astonished Sammy. When I climbed on to the seat of the plough, grasped the reins and screamed "giddup", Sammy hit the collar and led the team around the paddock for two full laps. Shortly afterwards I met one of those chaps who know everything, from whom I learned that I should have pinned the jib's ears together with a safety pin. According to him, trying to shake one ear without the other takes the animal's mind off jibbing. I have never had the need to try out his method, but I know from experience that my way worked.

It was while I was in this locality, which I think was named Numalla, that I had my first experience with girls. It was quite a depressing experience and may have had an influence on my attitude to women for quite a while. I had a contract to burn down and burn up old ring-barked trees in a one hundred acre paddock and was camped in an old one room shack, close to this little centre which consisted of not much more than a timber weatherboard community hall, where a dance was to be held. On the day of the dance, I decided I would go to it and started to make myself presentable. My hair was long and badly in need of cutting. Those were the days of short back and sides and, having scissors and hair clippers, I set to work.

It was a painful business and rather difficult, as I didn't have a mirror and had to improvise with the shiny base of a flat fifty cigarette tin, to reflect a rather murky picture of the back of my head on to a rather dirty window pane. The result was at best patchy, with numerous bald patches where the clippers had bitten too deeply and in other places some tufts that I had missed with the scissors. My suit of heavy English tweed probably made me appear a bit of a yokel, but I was well scrubbed and had washed a shirt for the occasion. I bought my ticket at the door in anticipation of an evening spent in the pleasant company of the opposite sex. As was the custom at such gatherings, the girls were seated along one wall, their mothers along the opposite wall, where they could keep a watchful eye on their daughters and monitor the behaviour of the young bloods who were gathered in a group by the door, peering out of a haze of cigarette smoke at the girls.

I knew no-one there, but quickly identified the best looking girl as the one sitting at the end of the row of chairs and next to the stage. I knew the protocol, as when I was about ten or eleven, I used to accompany my sister Rene to dancing lessons. I don't know whether my main function on those occasions was as a chaperone, an incubus to deter any forward youths from making a pass at Rene, or as a dancing partner, but whatever the reason, I had learned the basic steps, the fox trot, one step, Pride of Erin, maxina etc., and I had no doubt of my ability to steer my partner around the floor. Placing myself in front of this lass, I spoke the password, "May I have the pleasure of this dance?" There was a moment's hesitation before the reply which regretted that "she was engaged for this dance". I turned my attention to the next in line and received the same answer. Having gone to the trouble of cutting my hair and arraying myself in my best clothes for the occasion, I felt that I should keep on trying with the rest of the young ladies, with the result that I went the length of the hall, receiving a similar answer from each girl in turn. At that point I sadly took myself off to bed. It was a couple of years before I ventured again to offer myself as a dancing partner to a female I didn't know.

It was inevitable that I should make another move to join my brother in the west, which seemed to offer better opportunities than New South Wales for people aspiring to be farmers at minimal financial outlay. The move was expedited by the early death of my father on 3rd November 1927. Mother cabled the sad news, of which I already had a premonition. I was awoken one night by someone calling my name. There was urgency in the voice and I leapt out of bed calling out to my room mate, "what's wrong?" He flatly denied having called out and cursed me for wakening him. The date and time on the cable coincided precisely, after adjusting for longitude, with the time Dad died. That effectively put a stopper on plans for the family to come to Australia. It was a sad blow. My father and I were very close. He died three weeks after retiring from the police force. He had already received a retirement present from the members of his division, but after he died, they took up another collection for a posthumous presentation of a handsome gold half-hunter watch. It came into my possession and has been my talisman ever since.

Having experienced the risks of train travel on my first attempt to travel west, I booked a berth on the *MV Karoola* for my next try. Naturally I travelled third class and, to the best of my recollection, the fare was £11. I shared a four berth cabin with three old men, well maybe not so old on thinking back, say fifty-five to sixty years of age. They shared the distressing habit of getting terribly drunk before going to bed and then getting up during the night to spew into the cabin washbasin. The stench was unbearable and I spent most of the nights on the deck, trying to sleep in a steamer chair. It was wonderful to arrive at Fremantle; the year was 1928.

Frank was there to meet me and had found lodgings in a small bungalow on the north end of Beaufort Street Bridge. It wasn't the most salubrious locality and I imagine the continual passage of trains must have created a problem at night, but I have never had a problem sleeping and I can't recall any discomfort from that source. The room cost eight shillings a week, which was a bit pricey against the basic wage for a labourer. I found that I had very little affinity with my brother and teamed up with another chap, a newly arrived Englishman, who claimed to be a veterinary surgeon. In retrospect I see him as having a large component of the con man in his character. This was to be the first of a number of occasions when I formed a business relationship with someone who had a good idea and some relevant skills, but no money to put the idea into effect. I usually had some money, not a great deal, but on this occasion enough to buy a sulky and a very old grey mare, that on the road definitely turned out to be "not what she used to be". The veterinary supplies required by my new friend were minimal, a gag and a large quantity of Epsom salts. Stocked up with some stores and a few essential cooking utensils, a billy can and frying pan, a couple of enamel plates and mugs to satisfy our own needs, and a bag of chaff for the old mare, we set off.

It was a slow trip, governed by the problems of the hills and dales traversed by road in the Darling Ranges. On steep up-grades we would have to get out and walk, sometimes even assisting the old girl by pushing or alternatively, hanging on to a rope on the down-grades, to prevent the sulky from over-running her. We progressed at a rate of about sixteen kilometres a day. Neither of us were concerned at the slow progress. The scenery was new to us both and in its Western Australian way, delightful. I had not been in Australia long enough to have lost my yearning for the English landscape, but enjoyed the blackboys, red gum, jarrah and the very diverse understorey of flowering shrubs.

It took ten days to get to Goomalling, where my erstwhile friend decided to establish his veterinary business. One town was as good as another to me and I doubt if the old mare would have gone any further without a good, long rest. An arrangement was made for Bert, the vet, to buy the horse and sulky on deferred terms, paying when he had the money, which proved to be indefinitely postponed.

Jobs were fairly easy to come by. Wheat was a good price and previous seasons had been good. Until I found my feet I took a job with a local carrier, who among other activities sold firewood. The local wood, a type of acacia, when freshly cut smelled strongly of raspberry jam, and hence was named jam. When dry, which was what was required for firewood, it was devilish hard and I was pleased when my turn on one of the trucks came around.

Intermittently I worked on farms. The Western Australian farming community proved to be more enterprising than where I had worked in New South Wales. Tractors were just coming in and one farmer I worked for had a "Holt 22" tracked tractor. I believe that it was the forerunner of the Caterpillar series and was later called the D2. The Fordson wheeled tractor was in more general use. In some circumstances it was a treacherous machine to drive. If the towed implement, usually a plough, got stuck in a boggy patch, the tractor wheels kept churning away and would very quickly lift the front wheels off the ground. If the clutch was not immediately disengaged, it would continue lifting and finally rear over backwards, as very nearly happened to me the first time it occurred.

1928 proved to be a bad year. The winter rains failed and the pastures were soon eaten off. By the height of the summer, the sheep on the farm where I was working were slowly starving to death. Daily I would feed them with Meggits Nuts, linseed meal compressed into a nut, a highly nourishing but expensive food which could only be fed out at a rate which it was hoped would keep the majority of the flock alive. Those that foundered had to be killed and for several weeks I would kill about ten to twelve sheep a day. Eventually I lost that job, as the owner could no longer afford to pay me. Luckily the carrier was pleased to take me back on the payroll and I moved back into the town.

Towards the end of 1929 an opportunity arose in Goomalling to acquire a motor repair business, which was attractive to my current mate, Fred Gaskin, who was at that time a mechanic with a local business. Once again I found myself in the same situation of having some money and no relevant skills considering going into partnership with a friend who had the skills but no money. That alone should have put me off, as he was earning good money and spending it as quickly as he earned it. Our combined funds were insufficient to meet the purchase price and it was pointless even to consider it, when out of the blue a cable came from my mother, saying that father had told her I wished to buy a small business with a friend, but didn't have enough money and needed £200 to make up the purchase price. If I thought it was a good business, she would like to help. I was aware that after father died mother had resorted to "spiritualism" and found a medium whom she visited about every three months. This was the first tangible evidence that I had that mother had established communications with Dad, something which she often mentioned in her letters and of which I was very sceptical. A few years later, on a return visit to England, we discussed this matter at great length, and to my amazement she produced a series of school exercise books in which she kept an account of the seances she had attended with this medium. I was astonished to read details of my activities and, in some cases, of plans for intended action of which I had not told her, a point which she proved by producing all of the letters I had written to her during that period, which confirmed that point. It was uncanny and, although I had no personal wish to become involved, I accepted my mother's experiences as entirely valid.

On receipt of her cable I accepted the offer, the money was

transferred by cabled advice from London and the deal went through. A few weeks later, the bottom dropped out of the business overnight. The financial failures on Wall Street, which brought the international financial structure crashing down, found expression in our little community in a drop in the price of wheat, on which it depended, from almost five shillings per bushel to one shilling and eight pence. It took a while, a couple of months I suppose, to realise that the farmers, our customers, were broke and unable to pay the bills they ran up. Against my own background of frugality it was incomprehensible to me that people engaged in what for many years had been good, profitable business, would not have money in the bank to draw on and we continued to advance them credit for a couple of months, before it became obvious that they were broke. When we demanded cash it was astonishing how quickly business dried up. Within six months we too were broke. In my case £550 had gone down the drain. I never recovered the money owed me by those farmers, who had booked up substantial sums for petrol, oil and tyres. Subsequently they moved into more profitable times and many of them moved down to Perth and a leisurely retirement. None of them paid me the money owed to me. On the other hand, I was still paying some of my debts until 1935.

It was a very dull life. Takings virtually ceased and had it not been for the good services of the manager of the local branch of the Vacuum Oil Company, we would have fared ill. He helped in two ways. Whenever a truck of cased motor spirit or kerosene came in, those being the days when motor spirit, kerosene etc., was shipped from the US in four gallon tins, two to a wooden case, I would be invited in to unload the railway trucks and stack the contents. The stacks were built to a height that necessitated hoisting the cases to the limit of my fully extended arm. It was very good exercise and the pay for casual labour, two shillings an hour, very acceptable. The manager also ran a dance band. His name being Inkpen, as you might guess, the band was named "The Inkspots". On finding out that I played the violin, he took me in. He played the drums and the band, which consisted of piano, saxophone/trumpet, violin and drums, would go anywhere around the district to perform. My reward was £1 per night, literally per night, as most dances, or balls for special occasions, began at 8.00 p.m. and went on until 3.00 or 3.30 a.m. On an hourly basis it was just a little more than could be earned from the sweat of the brow. In combination, they kept the wolf from the door.

Not much motor repair work came our way, so that when a very travel worn woman of indeterminate age – she could have been anything from thirty-five to fifty-five, her features were well obscured by a heavy overlay of grime – came in late one afternoon and asked for some attention to her old Ford Flivver, it was no great work of charity on our part to attend to the repairs. It didn't really endear her to us to find that she had gone to each of the several motor repair shops before coming to us, after the others had refused her. She was on her way further up country, rabbiting, using ferrets, and all her worldly possessions were loaded on this old Ford car, in which a big end bearing had failed. She was of most unsavoury appearance, swathed in a long, threadbare, brown overcoat, with a trimming of ratty looking fur around the collar. Her face, filthy and lined, with a couple of teeth missing, was framed with long, lank, greasy hair and at the other end, her feet were encased in a pair of very old, much worn, down at heel, elastic-sided boots. I put it to my partner that it was only a matter of work, no money was involved in buying new parts, just take the sump off, disconnect the big end, run some metal into it, fit it and put the sump back on, and hey presto, a couple of hours would do the job. We finished the repair by about 8.30 p.m. and on handing it over, she was obviously quite distressed that she couldn't pay us for it. Taking me aside, she tearfully admitted that she couldn't pay, as she had no money.

"But I tell you what", she said, "you can have one of my ferrets". When I declined that offer, she made the supreme sacrifice, "I tell you what, I'll sleep with you if you like". I hastily backed off, telling her "I really hadn't started that sort of thing yet" and recommended she try my partner, on the grounds that he was quite a bit older than me and was used to it. It was an unkind thing to do, as she was sincere in her desire to pay and Fred's remarks, beginning with "Get out, you dirty old bitch" and very rapidly becoming more personal and abusive, were even to such a hardened character likely to be most hurtful. Despite her physical drawbacks, she must have done a little business after she left us, because first thing next morning, as she was heading north, she stopped in an unsuccessful endeavour to pay us a few shillings. Perhaps Fred's remarks had struck her pride. I prefer to think that although she was dreadfully dirty and had reached a rather low ebb, an honest heart beat beneath that grimy exterior.

Our landlord had the habit of coming round at the last day of every month, holding in a very trembling hand (I think the poor fellow must have had Parkinson's disease) a receipt already made out for the next month's rent. When in May of 1930 we were unable to dredge up enough money, he immediately served us notice to quit, and the sooner the better! This posed a problem. Having become proprietors of a small business, we didn't wish to relinquish that position and join the ranks of the unemployed, although after discounting the small effort involved in pumping petrol out of a bowser for sale on credit to farmers, who had no intention of paying and who played us for the suckers we were, we had been virtually unemployed for the previous few months.

At the back of the building an assortment of old automobile wrecks had accumulated over the years, one of which had aroused Fred's interest. It was what remained of a Dort car which had been completely burned out. It was the only Dort car that I have ever seen. I think it was of German make, probably manufactured in Dortmund. There were certainly no spare parts in Western Australia, even if we had the money to buy them. Fred had done a remarkable job getting it to go again. There were sufficient old tyres and tubes around, other people's cast offs, from which to select six of each to make the car mobile and have a couple of spares. The upholstery was all burned out, but we managed to scrounge and modify enough for the front seat and we cut the back of the body away to convert it into a utility.

It wasn't a very handsome vehicle. It had been standing in the weather for some years, which had generated a thick layer of rust on the metal that the fire had exposed, but it worked and we evolved plans for our future. A new railway had been constructed from Kalannie to Bonnie Rock, which had opened up some quite good land, but which was subject to an uncertain, highly variable rainfall. Part of this land had been made available to "dusted" miners, disabled as a result of inhaling silica dust in poorly ventilated underground mines, and to ex-soldiers for settlement and two or three towns had been gazetted. We decided to try our luck at one of these towns, Beacon, and get in on the ground floor. The Dort would get us there.

Before we left, we decided to make a concerted effort to try to get some money in from our debtors. Having partitioned the district between us, and carefully tabulated those who owed us the most, and those who were most likely to pay, we set out early one morning, each to his zone, to see how much we could get in. When we returned at the end of an arduous day, and compared notes, we had been equally unsuccessful. Neither of us had collected a cent, or in terms of the day, a penny. We sat there very glum, when suddenly Fred said "Kemp! We haven't done Kemp yet". Kemp had a small shop a few doors down the main street from our location from which he sold vegetables. He was an active, skinny little fellow, but no-one carried much weight in those days, with a Polynesian wife and two quite attractive daughters, whom he was doing his best to support.

When we presented our account and suggested he pay it or

pay something off, his face dropped and his halting explanation why he couldn't pay brought us the recognition that we were on to another loser. He had no money. Fred was a resourceful fellow and, turning to leave the shop, he stopped and said, "Have you got a sack, Kempy?" "Yes, yes, sure, I'll get you one". On receiving it, Fred began to fill it with vegetables from the shelves, which reduced Kemp to tears, complaining that if we took his stock, he would have nothing to sell to get money to buy more and his business would fold. What could a person do in such circumstances? Reluctantly we replaced the assorted vegies on the shelves, retaining only three or four large ironbark pumpkins. They were to play a significant part in my diet for the next couple of weeks.

The following morning we drained the residual petrol from the bowser into the Dort and some twelve gallon drums, loaded the transportable tools, our bedding, clothes, eating and cooking utensils and Fred's old Harley Davidson motor bike, and took off for Beacon. It was an eventful trip. The Dort had a major problem, the vacuum tank. Those early model cars didn't have petrol pumps to feed the carburettor, it was sucked up from the main tank by a vacuum, created by the action of the engine, into a vacuum tank, which from memory would hold about a pint of petrol, recharged at a rate commensurate with the rate at which it was consumed.

This was the first (and only) time the Dort had a long and continuous run and it revealed that the vacuum tank leaked air through a very thin gasket, which should have sealed the removable head to the tank.

The condition caused some delay, but we eventually got away and completed the two hundred kilometre trip by providing an artificial source of vacuum. There was a curved, small diameter pipe, say three or four millimetres, extending from the vacuum tank to void surplus fuel if the tank overfilled. When the tank was empty, by prolonged and vigorous sucking, it was possible to recharge it. We proceeded with me sitting on the left hand mudguard, from which position I would regularly bow my head, take the pipe firmly in mouth and suck. When I got a mouthful of petrol, I knew the tank was full. We proceeded in this fashion for the full distance, stopping only to mend numerous punctures. As may be readily imagined, we got our full share of attention at every town we passed through. Most of the roads we travelled were gravel and all badly corrugated and the trip was filthy, uncomfortable and precarious, for there was little solid support to hang on to. But we made it before night set in. Fortunately the weather was fine, as until we erected some form of shelter, we were exposed to the elements.

The following day Fred set off for Perth on the Harley Davidson to see if he could get any form of assistance, either from the bank or suppliers, to get us going again. We were already in debt to most of the automotive suppliers and, considering the limited activity in the district and the impoverishment of the settlers, it is small wonder that when he came back, he was empty-handed. He was away ten days, during which my task was to erect the framework of a bush motor repair shop. We had carried up enough old, well-used corrugated iron to roof a small building, but very little else. The small amount of cash we had, Fred needed to see him through his trip to Perth and back. I was left with the pumpkins which were my sole item of food, until, on his return, we decided to risk running up an account at the store, which was at the time the only other building at Beacon. For a few days I had the residue of the tea and sugar etc., we brought from Goomalling. After that it was pumpkin cooked in a few different ways, boiled, baked in ashes, fried in fat, washed down with water.

It speaks volumes in favour of a vegetarian diet that I was able to proceed with the construction of the shed, an arduous task, consisting of cutting good, straight poles from the forest, twelve feet long, with an average diameter of eight inches, carrying them on my shoulder back to the construction site and embedding them three feet deep at intervals around the perimeter of the proposed building. Fortunately the forest was undisturbed right up to the railway reserve, which in those days usually formed one boundary to the town and fronted onto the main street. Not all the suitable poles, of which I needed about twenty, were close to the site. On average they had to be carried about three hundred yards. The transportation procedure, after trimming and debarking the log, was to raise it at the small end and walk along it until the point of balance was reached, smartly get a shoulder under it, adjust the balance and then run like hell until the pain became unbearable. A rest was then taken until the rubberiness had gone out of my legs and the pain out of my shoulder. The procedure was then repeated until home base was reached. A stick of hardwood of those dimensions, cut green, would weigh anything up to three hundred pounds, and thirty yards at a time was about as much as I could manage. In these circumstances there was no room for an eight hour day or weekends but by the time Fred came back, I had the framework of the building up and the service pit dug and framed. Between us we got some smaller diameter sticks up for roof support, the roofing iron secured and strips of hessian stretched around the frame for walls. With a chain block suspended from a reinforced rafter in the roof, our tools and garage jack ready to hand, we too were ready for work.

Unfortunately none came. There was no passing traffic and the few locals were apparently self-sufficient. One small job came our way early on to break the monotony and provide a small amount of cash, when one of the miner settlers, Jim Chesson ex Cue, wanted a one thousand gallon rain water tank assembled from the component sections of pre-curved corrugated iron supplied to him. It was a plumber's job, a trade in which Fred had neither training nor experience, but as we were quite broke, we decided to have a go. The best that I can say about it is that when finished it was watertight, but its shape was far from regular and the owner far from pleased. However, he paid up and progressively a few more jobs trickled in. We became known in the district and the farmers, bless them, although they had very little themselves to give away, would assist with the occasional gift of wheat or milk, or even meat when they killed a sheep for their home consumption.

The days were long and boring, without the relief of some satisfying work. The evenings and indeed some of the nights, were mostly spent playing bridge. Occasionally a farmer would invite me and a few others out to his place for the evening meal, followed by some light entertainment to which everyone would contribute according to individual talents. Nearly every family in those days had an upright piano, usually badly out of tune, which one or other of the family could play. I would produce my violin, almost everyone present could sing, a song book would be found containing the old imperishable favourites, both for solo voices and part singing, and for several hours "the cares of the day were charmed away", as the song "My Little Grey Home in the West" has it.

We failed in our attempt to make a real business out of it. There simply wasn't work around and I applied for and got the Wesfarmers' contract for the wheat stack at Marindo, six miles west of Beacon. Fred departed to seek work elsewhere. In those days wheat lumping was laborious work, consisting of stacking the wheat in layers, twenty layers high with the bags laid in the pattern of Flemish Bond used in brick laying. Some degree of skill was required to ensure the stack didn't fall over.

The wheat bags, made of jute and manufactured in India, were designed to hold three bushels when full and at sixty pounds a bushel, weighed one hundred and eighty pounds. The bags were a significant cost element in marketing wheat off the farm and, with the very depressed price of 1/8d a bushel, the farmers resorted to slitting the triple hem at the top of the bag, around which the bag was sewn together after filling, as by this means an extra twenty pounds or so of wheat could go into the bag. In some cases, by strongly ramming the bag, some farmers could get two hundred and ten pounds into a bag. This posed a real problem for the lumper. A bag which was normally filled to one hundred and eighty pounds could be tipped off the shoulders from a height of four to five feet, but with the weakened hem and the bag rammed to two hundred pounds plus, it was almost necessary to kneel down and let it roll off gently, to prevent it bursting open at the stitching.

A farmer would normally bring in twenty-four bags on his farm vehicle. Each bag was individually weighed on a manually operated scale, the weight recorded on a tally sheet and written on the bag, using a split cane which had been dipped in blue dye. It was such an archaic system that the dye could credibly have been woad. To enable the lumper to build the stack twenty bags high, two oregon planks about twenty feet long, twelve inches wide and two inches thick were used. Initially, after those that could be stacked from ground level had been placed, one plank went from ground level to a position on the stack about five feet above ground. As the stack grew in height, a second plank was used and a wooden stand introduced into the system to support the elevated end of the first plank and the bottom end of the second, the other end of which would then rest on the stack, about ten feet above ground level. After weighing, recording the weight and marking the bag, the lumper would then take the bag on his shoulders and run up the planks, which were springy and responded to the rhythm of the lumper's steps. After reaching the stack, he then walked and climbed to the position at which the bag was to be placed. It was hot and heavy work, paid on the basis of two pence per bag from the farmer's truck into the rail truck, whether directly into the rail truck or via the stack, in which case the lumper was paid one penny a bag into the stack and one penny a bag from the stack into the rail truck.

With luck, a rake of railway trucks would be dropped into the siding at the height of the harvest season (a rare occurrence) and it was possible to receive and load one thousand bags in a day, which at two pence a bag was worth £8/3/4, but well over three-quarters of the bags went into the stack at a penny a bag. To handle six hundred bags a day, with a large proportion of them being carried to the top of the stack, was a heavy day. Good legs were needed for this work. The first trucks would start to arrive at about 6.00 a.m. and carting would continue until 6.00 p.m. There were, of course, no restaurants to eat in, or wives to prepare meals and most meals came out of tins.

The work of receiving the wheat, stacking it and subsequently outloading it was contracted out to individual lumpers or, if the amount coming into the siding was very large, to a party. On average a two person stack would be for sixty thousand to eighty-thousand bags equivalent to £500 to £700 spread over eight months starting slowly in November, building up to a peak in December - January, fading in February by which time all of the wheat had been delivered. During the time that deliveries were trailing off, the stack would be roofed with corrugated iron using an ingenious method whereby the iron, sandwiched between wooden laths, would be held firmly in place by the weight of the underlying wheat bags. Outloading would then commence, the rate depending on the number of rail trucks delivered daily at the siding.

Chapter 3

It was quite fortuitous that I became involved in the mining industry and a further confirmation of my belief that everything turns out for the best. The failure of the small business venture at Goomalling seemed, at the time, to be a terrible blow of fate from which there was no early recovery. Some of the additional money I made from wheat lumping, I used to pay some of the most pressing debts and retained a small surplus, which I allowed to accumulate with the grain merchant. Regrettably for me the Trades Protection Association became aware that I had a small credit balance with Wesfarmers, against which they lodged a garnishee, I think the amount was £25.

Fortunately I had a credit also of £25 with Bunge, another grain merchant, payment for salvaging a wheat stack that someone else had built badly and which had fallen down. I lost no time in cashing this and ensuring that I retained ownership by wearing it in a money belt next to my skin.

The time ran round to August/September 1931. Wheat lumping was finished for that year and I spent several weeks sitting it out in my camp in Beacon, waiting, like Mr Micawber, for something to turn up. Something did, but of no great moment, other than to illustrate the state of the times. One of the farmers, who had been bringing in wheat to my siding, asked me if I would be interested in helping him put up a fence on his property. He was rather reluctant to be specific regarding wages, preferring to soften me up by saying he didn't expect me to accept, as he could barely afford to pay me anything at all. Eventually he disclosed that he could only manage ten shillings a week and keep, and assured me that I would not be called upon to do any domestic chores, such as milking or chopping house wood. He was a very persistent chap and after telling him on several occasions that there was no way that I would accept such miserable conditions, I capitulated.

Besides being an excellent farmer, he was a good judge of character. On the first morning at work, as the sun was appearing over the horizon, he lit a small fire at the end of the fence line in a token effort to take the frost off the crowbars and with minimum delay, we set to. He had a good fund of stories, mainly about his own experiences in the bush, but he kept his bar dropping at a good round pace and I was rather disconcerted to find that he had started on his second hole before I had finished ramming my first post. I smartened up my pace, he increased his and within the hour we were both working away as though being paid contract rates. The outcome was that we finished the job, which under normal conditions would have taken four to five weeks, in two weeks. My remuneration was £1. However, he had been as good as his word. His wife was an excellent cook and they fed me very well for the prevailing conditions. I had previously been batching at an all-up cost of five shillings a week with boiled wheat and treacle being a substantial part of the diet, but with the menu helped out by the occasional gift of a few eggs and some milk from some of the farmers' wives.

These conditions were not conducive to sitting around waiting for more of the same and I decided that I would roll up my swag and make for Geraldton, to see whether I could contract from Wesfarmers the stack at Ardingley, a siding six miles west of Mullewa. I had worked on the stack two years before with a couple of chaps from Fremantle who had the contract and paid me wages, plus a margin. I was accompanied in this venture by Joe Bell, a chap of about my age, who professed a willingness to throw his lot in with mine. After storing belongings surplus to what we could comfortably carry on our backs, we took off.

Although it was well into spring, when the sun went down on the first day out, there was a decided nip in the air, which held promise of a cold night. As we only had five blankets between us I told Joe I thought we should roll up together, getting the benefit of our mutual body heat and the extra blankets. Joe shied away from this like a nervous colt. The poor fellow must have had some unpleasant experiences in similar circumstances before. Anyhow, I could see no practical reason why he should object to my proposal, since sodomy was not in my repertoire and in any event we would be sleeping fully clothed. I told him that I proposed to keep three of the blankets and that he would have to make do with two. At some time Joe was probably a boy scout or perhaps a cub and had learned how to make a comfortable bed by building a mattress from leafy branches and this he proceeded to do. I scraped a hole in the loose loam which took me down to below the wind level, rolled up in my three blankets and rapidly fell asleep, but not for long. It was probably two hours later that Joe, with his two blankets, was pleading to be allowed to join me. The cold wind drifting up through his carefully constructed and elevated perch had proved too much for him and the poor fellow was half frozen. That first night apparently put his fears to rest.

When we reached the main north highway at Pithara, we didn't have long to wait before we were offered a lift by a youngish woman going to Dalwallinu. There was none of the apprehension in those days which can be legitimately entertained now, that offering a ride to someone travelling on foot can be a hazardous business. Shortly after she dropped us off, a heavily laden two and a half ton Dodge truck pulled up and took us on board. After enquiring where we were going, the driver strongly recommended that we should accompany him to Meekatharra and then jump the train down to Geraldton. To do so would add a few hundred miles to our journey, but it would eliminate the walking. After pondering the idea for a few minutes – I didn't have much time to consider it, as he was leaving our planned track a few miles up the road at Wubin – I decided in favour of it and putting my head out of the cab, informed Joe, who was riding on top of the load, accordingly. He was not much taken with the idea, but liked the alternative of proceeding alone even less.

The driver, whose name turned out to be Baumgarten, was an interesting chap, the sort of fellow who spent all his life in the bush, except for the occasional visit to Perth for a rip-roaring burst of riotous living. It turned out that he was just returning from such an occasion. He was a member of a party working a small gold show in the Nabberu area, about one hundred and eighty miles north of Meekatharra. He had been to Perth to get some much needed stores and equipment and wasn't looking forward to the meeting on his return, as the other members of the party had given him the last of their combined cash to buy tobacco and having blown the money on a good time, he was returning empty-handed. Quite rightly he was fearing the consequences. For a smoker, living a spartan life in the bush, tobacco has a special significance. They had used the last of the tobacco before he left and I could imagine them counting the days off until his return. Perhaps it was because he felt that he needed some insulation at the reunion, that he told me with some fervour that he wished I was going the rest of the way with him.

His conversation during the run from Wubin to Meekatharra was mainly to do with prospecting and gold discoveries and it had stirred my interest. We overnighted at Ningham and passed through Mount Magnet and Cue, two gold mining ghost towns, next morning. Both caught my imagination and I liked the country, even though it was flat and dry and barren, but when we pulled into Meekatharra at midday and drove down the wide, gravel street, lined with its wild western-type shops, something in my mind clicked and I decided on the instant that it looked like the threshold of a new and different life, rich and exciting, that I would like.

I told Baumgarten that I would go on with him, broke the news to Joe and put up with his complaining until he had got used to the idea. We stayed in Meekatharra the rest of the afternoon, during which I unloosed the money belt and prised a pound note out, with which I bought fifteen two ounce tins of Westward Ho! tobacco at one shilling and four pence a tin and with some loose change I bought some Rizla cigarette papers. Without telling Baumgarten or Joe what I had done, I rolled them into my swag. The track to the mine, and it was no more than a track, was in many places in a terrible condition and we drove all through the night to get to the camp before nightfall the next day. The chaps had heard us coming and had run a half mile down the track to meet us, with cries of "tobacco, tobacco, where is the tobacco?" When the dreadful truth was out, that there was no tobacco and no apparent chance of getting any for a few months ahead, their wrath was turned on to me and Joe. "Who are these bastards anyway?" they snarled. It was their main theme, with many colourful variations. Before they strung us up to the nearest tree just to ease their feelings, I unrolled my swag and with a few conciliatory remarks about knowing how hard it was for a man to be deprived of a smoke, I presented them with "the makings".

The gift wrought magic. I was their friend for life. The joy of the besieged at Mafeking or Lucknow on being relieved was nothing compared to the relief of those chaps. We stayed with them for four days, during which time they persuaded me to join their party. This involved going back to Perth, getting a vehicle, stocking up with stores and importantly, the gear that was available from the Mines Department to equip the hopeful prospector, comprising of a windlass barrel and stand, a length of wire rope, two safety hooks, two hide buckets, a gympie hammer and a set of drill steel, a miner's pick, a couple of shovels and two gold dishes.

The trip back was rather more arduous than the ride in. From the camp to Peak Hill was about one hundred and twenty miles and had to be done the hard way on foot. They stocked us up with food, but it was very rough tucker, mainly home-made bread, made with home-made yeast, which had gone sour making the bread almost inedible, some kangaroo and fortunately, some flour. After the first attempt at the bread, we impaled it on a dead sapling on the roadside, where it defied the bird life and the weather for many months. It had been a good winter and there were enough pools in the creek beds to ensure our water needs. The weather was fine and we moved along at a good pace, encouraged by the thought of purchasing some food, notably bread, meat and sugar at Peak Hill. However, it was not to be. The storekeeper, a man called Stalley, proved to be a very rough character. Perhaps he had had a row with his wife that morning, or maybe he had a bad experience with other chaps reduced to carrying their swags, but he refused to sell us anything, abused us roundly and smartly got us out of his shop. There were no other stores in Peak Hill and we had no alternative but to resume walking towards Meekatharra.

Walking on an empty stomach is hard going and we were pleased when we reached Bilyuin Pool and the wayside hotel, about twenty miles down the road. The pool had a fair amount of water in it and a bush fishing pole, complete with hook and line, was leaning against a shady tree that overhung the pool. Leaving Joe to try his hand at fishing, I went to the hotel to try my hand at getting some food.

It was Sunday afternoon and the front door was firmly closed, but I could hear the sound of a piano being played. In response to my knock, the playing stopped and a grey-haired woman, about fifty or so, appeared at the door. Almost before I had voiced my need for victuals, she grabbed me by the arm and pulled me inside, whilst telling me she hoped that I liked music, because she had just composed a piece and wanted to play it to me, to see if I could recognise her source of inspiration. Placing me comfortably on a sofa, she had just seated herself at the piano, when an elderly man, who I assumed was her husband, came in and asked me if I would like a beer. After just having walked twenty miles on a hot day, I certainly did, but told him that I didn't have the money to buy one. "That's alright mate, I own the place, have it on the house", he replied and offered me a cigarette. Within a minute of my arrival, I was puffing away, sipping at a sixteen ounce pot of cold beer and listening to Mama playing her own composition, which I was certain was intended to represent a magpie. When she asked me to identify the divine source of her music, I equivocated, telling her that I recognised the pastoral quality of the music, rippling water, bird song . . . "It was a magpie", she burst out triumphantly.

Crediting me with successfully recognising the source of her inspiration, it was impossible to get away. The old man kept bringing in more beer. His wife kept playing more of her compositions and a good two hours passed before I was able to explain that I had a mate waiting for me at the pool and that my main mission had been to see if I could buy some bread, meat and sugar. "Buy some food? Unheard of." Within moments I had a sizeable piece of goat, a loaf of bread, sugar and a bottle of goat's milk pressed into my hands.

It was these that saved me from Joe's wrath. He had caught no fish and was hopping mad. The food, barbecued goat and bread, washed down with tea made with milk and sugar, restored him to his normal good humour. We rolled our swags out and had just turned in, when a lad of about fifteen walked into the camp to invite us to go back to the pub when we had eaten, for a sing song. Regrettably we had to disappoint them. There were still two days of hard walking between us and Meekatharra and we needed an early start next day. The Garlands were a nice old couple. He worked a small gold show during the week and she ran the hotel. Not much traffic passed that way at that time and I guess it was subsistence living, but they had a full measure of warm-heartedness which is usually found in people living in lonely places.

We reached Meekatharra two days later after having been picked up by the local water supply officer in his utility. Following the general custom, we camped in the shelter at the recreation ground. The next morning we divided our food gathering chores, with me going to the butcher and Joe taking on the baker. Meekatharra lived up to my initial appreciation. On asking the butcher if I could do some work on his wood heap for some meat, he replied that the wood heap was private, "that's where I get my exercise," but when I turned to walk out, he called me back. "Where are you going? How many of you are there?" and bringing a slab of rump steak out of his cold room, he cut off two thick slices, which he gave me with his best wishes for our future good luck. Joe, who couldn't brace himself to go into the baker's and make a similar request, fell under the gaze of the baker as he passed and repassed the shop. "Is it a loaf of bread you be wanting, my lad?" The baker was an Irishman. "Be you alone or have you a mate? Shure 'tis hard times my boy, but take this and good luck", as he gave him a double sandwich loaf.

The water supply officer had a job to do at Nannine, about thirty miles on our way and took us with him. Shortly after he dropped us off, a freight train pulled up to take on water and as it was heading west, Joe and I scrambled up between two sheep trucks. These are constructed with a four inch footing on the end of each truck and with one foot planted on each, it was possible to get a fairly secure ride. However, the tracks had never been properly ballasted and when the train got up a bit of speed, it swayed rather violently. It was about 4.00 p.m. when we left Nannine and by the time we reached Mount Magnet, it was about midnight. The night had turned cold and Joe kept saying that we couldn't go much further. He passed over his swag and the water bag for me to carry, in addition to my own. What with the endless ribbon of bright steel passing below and the cold and the additional gear, I was only too pleased to get off myself.

The comparison between the reception we received at Mount Magnet and at Meekatharra was interesting. I was fobbed off by the butcher, whose shop was festooned with sides of mutton and pork, half carcases of beef and long strings of sausages. "I don't know what I'll have left", he said, as though I was asking for a couple of side of sheep, "come back later". By this time I knew his type and decided against going back. Joe worked all day, mucking out the baker's stable and yards for one loaf of bread. The next morning we once again took off on foot and an empty stomach. By nightfall we had knocked off twenty-one miles to reach Yoweragabbie, a sheep station, the name of which had been transferred to the railway siding. The lady of the house redirected me to the shearers' quarters, shearing being in progress. It was a Saturday afternoon and only the cook was in evidence. He attended to our immediate needs, with tea and brownie, and when the team returned from the town pointed out the shearers' foreman to us. He was very positive. "Roll your swags out in the bunkhouse. You'll find a couple of spare places there and every time you hear the tucker bell ring, line up with the rest." We feasted royally on Sunday and after breakfast on Monday morning when an early train was due, they turned out in force, flagged it down, saw us safely into an empty truck and on our way to Geraldton.

We stayed at Geraldton a couple of days, not from choice, but in considering the best way to get out and on our way as quickly and inexpensively as possible. Geraldton was the terminus for the Midland Railway Company, which was owned by an English company that had been given extensive land concessions in return for constructing a railway line from Midland Junction to Geraldton, which ran about midway between the Government line and the coast, and then parallel with the coast from Dongara to Geraldton. The Government rail service didn't have a policy which required their staff to prevent people from hitching rides, assuming, probably correctly, that they were travelling from one place to another in search of work. The Midland Railway Company were a different kettle of fish, very strict and not averse to laying charges against anyone unlucky enough to get caught. Reconnaissance of the Midland yards indicated very clearly that the prospect of hitching a ride right from the terminal was impractical, so reluctantly I purchased two third class fares to a little place called Walkaway. It almost proved to be aptly named. Alighting from our carriage, we doubled back and scrambled into a truck, partially loaded with wool bales, which were covered with a tarpaulin and resembled a tent. We lay doggo for some time, during which there was much to-ing and fro-ing, after which the engine gave a triumphant toot and left our truck stationary on a side line. A few hours later, another freight train came through and we managed to find a seat on the tender, much to the surprise of the guard when he later discovered us and threw us off. "You can't ride there", he said, "get off and beat it. This is private property". "But we can", I said. "We have been riding here since Walkaway". "Don't give me none of your lip. Just get out of here quick". But it was dark, and although he tried his best to prevent us getting on to the train somewhere along its length, it was a vain task and we were carried along to Mingenew, where we dropped off for the rest of the night. Rail traffic was not very frequent on that line and it was two days before another freight train came through, but it included the wool truck that we had abandoned three nights before. The guard on this train had a kindly blind eye and carried us in great comfort, almost into Midland Junction. We had the sense to drop off before it entered the terminal. Getting back to Beacon was comparatively easy. There was much more traffic and we covered the final stretch in three days, completing about twelve hundred miles in thirty-five days.

☆ ☆ ☆

I had made up my mind to return to Nabberu and Joe decided otherwise. Now having a destination and a job at the other end, even though I did not know how much if anything at all it would bring me, I decided I could spend some of my money. I had a few personal items to get down to Perth, where I intended to recruit a partner. These included some well-loved books and my violin. As I could not conveniently carry these travelling on foot, I took the train. A week's lodging cost me five shillings and I ate at a cafe opposite the Wellington Street railway station, named the Golden Eagle, where meals cost six pence or ten shillings a week if the full twenty-one meals were taken. The place was clean and the meals satisfying and tasty. I can't imagine how they did it, yet I am sure the proprietors made a profit. Just think, five cents for a three course meal and compare it with today's prices, when a piece of steak at a restaurant costs from $15 to $25 and vegetables are extra! I advertised in the *The West Australian* for a mate with a vehicle and enough money to pay for his share of the stores we would need. I eventually took in two chaps, Colin Blakeway and John Stewart, who were mates. One of them had a very old vehicle, a Hupmobile, which had been cut down from a touring car and converted into a utility. Between them they could muster a bit over £10. I made it up to £20 and with that we bought good hard rations, flour, rice, ironbark pumpkin, dried beans, some tinned meat for emergencies, a couple of sides of bacon and tea, sugar, dried fruit. We had a .303 army rifle and one hundred rounds of ammunition. I also had a .32 and Colin a .22 with which we expected to keep ourselves in fresh meat. We also had extras like jam and treacle to help the home-made bread and damper go down. A drum of petrol completed the load.

As my footwear was in very poor condition and would not stand up to much walking over rough ground I bought a pair of working boots. Not having much money to spend lavishly I bought a pair of Blucher boots, no doubt named after the German general who helped Wellington beat Napoleon at Waterloo. The uppers were cut from one piece of undressed leather which was fastened to the soles with short brass screws and the boots cost five shillings. Regretfully they did not last very long in service as the temperature in the unventilated mine was very high and the perspiration trickling down my body finished up in the boots, which then emitted sloshing noises with every movement and quickly rotted the soles and uppers which separated. I improvised a pair of sandals from the outer cover of a rejected motor car tyre. Cut to the pattern of the foot and tied on with strips of kangaroo hide they were satisfactory for walking around in but didn't offer any protection from falling bodies. It was just as well that there was no Occupational Health and Safety Officer around in those days.

The trip up to the mine was a foretaste of things to come. The tyres were in poor shape from standing in the sun too long. In any event the quality of the rubber was vastly inferior to today's product and even with new tyres on a made road, patching leaks was a frequent occurrence. The patching procedure was novel. It consisted of clamping a raw rubber patch, which had an incendiary backing, over the leak in the tube and igniting it, which then cooked the rubber over the hole. It was a slow and laborious task, taking the wheel off, removing the inner tube from the tyre, patching it and then reinflating it with a hand pump. When performed perhaps a dozen times on a hot day on a dusty road, it was positively irksome. We used many tins of Schrader patches on the trip, but, although I had doubts of the ability of the Hup to cope with the very many steep-sided, sandy bottomed creeks we had to cross, we made it. On our arrival the "miners in residence", as it were, decided they were due for a trip to Perth and after giving us a brief lesson in the use of explosives, left us in charge.

At the time they were deepening the shaft from the ninety

feet stoping level, to a level to be established at one hundred and twenty feet below the surface. Fortunately the ground was well-oxidised, easy to drill and broke well, so that despite our lack of experience, we pushed it down before the party arrived back. It was only a small hole, about four feet by two feet, and the first time I descended on the bosun's chair, I jammed about half way down in a tight part of the shaft, firmly held by my knees on one wall and my buttocks on the other, and the slack rope paying out below me. I don't know what the chap on the windlass was thinking about, not to have noticed that he had lost his load, but I guess it was either women or beer!

When the main party returned, we started knocking out a crushing in earnest. It was (we believed) high grade ore: it had to be to make it worthwhile carting one hundred and twenty miles to the State Battery at Peak Hill. We were aiming for twenty-five tons, which we expected would average four ounces to the ton, which at £4 an ounce, would have grossed £400. The proceeds had to be split between seven, the four in the original party and we three new ones. Not everyone worked in the mine. Some were out prospecting and one did the camp chores, mainly cooking. Working underground was the best of these jobs, as the mine was at a constant moderate temperature, whereas it was invariably stinking hot above ground. The main drawback to underground was the poor ventilation, just the one small shaft with the workings off, with fresh air introduced through a four inch pipe, fed by a wind sail. Unfortunately, the wind didn't always blow and the flame of the candle, our only source of illumination, would often get very weak. It was then time to get hoisted to the surface and have the windlass man introduce some fresh air into the workings by running the bucket up and down the shaft at high speed for quarter of an hour or so.

Living in close daily association with a small group of people for a protracted time without the relief of external contacts can certainly test one's patience and sense of humour and tends to develop a high degree of intolerance to the other person's sensitivities. On one occasion two of the chaps had gone out early one morning hoping to shoot a kangaroo as we were low on meat. They returned late in the evening in very bad humour and empty-handed. While they were away I had produced a culinary special out of some meat of doubtful quality, a few potatoes and onions, both of which had gone soft, and had thrown out a good crop of shoots. I finished off this dish, which could pass under the name of casserole, by make a crust of mashed potatoes tastefully ornamented with designs traced on with a fork, finishing it in the camp oven and then disguising the flavour with a liberal application of pepper. When I served this gastronomical masterpiece, carefully preserving the artistry on the crust, one of the chaps raked it over with his fork and in a voice designed to get an unfavourable reaction said "What's this shit?" That remark established in my mind for all time an awareness of how sensitive people can be to their cooking and ever after, whether eating at home or away, I exhibit signs of great relish regardless of whether I liked it or not and, if the cook is a novice, say what a wonderful dish was served up.

After a couple of months, the tinned meat we had included in the food we brought in was all eaten. Our expectation that we would maintain a meat supply by shooting kangaroos was never realised. They would come in droves of two hundred to three hundred, stay for a day or two and then move on, not to be seen again for about a month. They could hardly be blamed for that, as the feed was practically non-existent. They were in very poor condition, riddled with worms, and anything but appetising. When they were around, we would invariably shoot a couple and, to make the meat last, we tried making biltong. The climate was certainly right for the process, very hot and very dry, and once we had learned to protect the drying meat from blowflies and crows, by hanging it up to dry in an improvised chaff bag meat safe, we could extend the use of the kangaroo meat for an extra couple of days.

It was a very meat-sparse diet and we developed a real craving for a good meal of meat, the effect of which led up to an event which in different circumstances might have enrolled us in the ranks of the criminal classes. We had been cleaning out an old shaft and progressively dug through successive layers of soil and hide, horns and bones. It was obvious that other people at this site, before us, had been living off the land on a more nourishing diet than kangaroo. A couple of the chaps must have turned over in their minds the thought that we were not showing the adventurous spirit, the "to hell with the consequences" attitude of those who had preceded us, because a couple of days later they started an insidious propaganda campaign. When we were sitting down to one of our normal unappetising meatless meals, one of them volunteered the comment, "By cripes, a nice fat juicy steak done medium rare would be a bit of all right, eh? Help to get this bloody pumpkin down, wouldn't it, eh? Yeah, a nice bit of rump cut thick and grilled over the coals!"

Before he expanded further on this mouth-watering, gut teasing theme, his mate weighed in with the observation that he happened to be taking a bit of a stroll down by the pool, when he noticed a small bunch of cattle. "They were none too fat, mind you, but there was a young heifer among them that wasn't too bad. Must be a better do-er than the rest, or perhaps she is still on her mum. Yeah, she didn't look too bad", he said. "She'd cut up into a few good steaks. You know when I spotted her I could almost smell a roast of beef cooking in the camp oven, with some of that bloody pumpkin and a few of those potatoes that have started to sprout." Human nature can only stand a certain amount of this sort of pressure and then it starts to crack. After a few days, the suggestion that we should knock off the heifer was unanimously accepted and done that very same day.

The seven of us, Baumgarten, van Wayland the Yank, Bill and Jim Donohue, Colin, John and I went out, located the straggly little mob and identified the beast marked out for slaughter. Considering the enthusiasm, the pitch of excitement that by this time had been worked up, I thought that everyone would be itching to grab the rifle. Then Yank, who was always a bit of a bush lawyer, said that knocking off a bloke's cattle was a capital offence. "If you get caught, they can hang you", he said. "Aw, come off it", said another. "The worst you'd get would be life". This caused the best shot in the party to discover that he'd got something in his eye and couldn't see properly, another chap had strained his arm that afternoon and as a result his hand was shaky, two others claimed they were such bad shots that they couldn't hit a haystack from a hundred yards.

In disgust at their pussyfooting, and urged on by the others, I took the rifle and with the shadow of the noose mentally hanging over my head, pulled the trigger. We bled the beast and dropped its innards out at the scene of the crime after which it wasn't much of a job to lift it on to the Hupmobile for transport back to the camp where we had rigged up a tripod, and got the skin off. Within the hour we had the beast hanging from the tripod waiting for the cool of the night to set it sufficiently for cutting up. Whilst we were contemplating this mouth-watering spectacle, one chap happened to turn and glance down the road. Followed a gasp, "For Christ's sake, here's Monty Bender coming with a couple of his blokes". Sure enough about a quarter mile away, three horsemen, one of them recognisably the owner of the animal that we had just knocked off, were jogging slowly into camp. There was a hurried, panic-stricken conference, which ended with me being shoved forward by the others to do the talking. I remember that I stumbled through an explanation of how hard we had been working, a statement that we could substantiate by pointing out the size of the mullock dump from the shaft we were sinking, that we had "just killed one of his cattle", a fatuous remark, because the carcase hanging from the tripod and gleaming redly in the rays of the setting sun, was the most conspicuous object in the landscape, that we hadn't had any meat for a long time, and ended up by saying "but you can have half of it if you like and – er – the hide". The tanned and weatherbeaten old face remained deadpan, there was a non-committal sort of a grunt and then, "Do you mind if we camp here tonight?"

The party rode off in the morning, leaving us in full possession of the carcase. It turned out to be an ill-fated business. We gorged ourselves on the meat and paid the penalty of eating too much while it was too fresh and all of us suffered diarrhoea for the next three days. We decided to salt the remainder, but none of us knew the proper procedure. We put so much salt into it that no matter how long we soaked it for afterwards, we could not get rid of it. I sadly recall that we ended up by throwing much of it away.

☆ ☆ ☆

I was sitting at the shaft collar one morning on duty as windlass man with my ear tuned to the four inch ventilation pipe which, besides being our means of ventilating the mine, acted also as an excellent speaking tube. I was listening to the odd snatches of conversation between the Yank and Bill Donahue working below and putting a word of my own in now and again, when I heard a cr-r-rump and felt a small rush of air up the shaft followed by an anguished cry of "Oh-h-h Bill!" It was from the Yank, van Wayland, who was working on the thirty metre level knocking out the reef. One part of this must have been holding a large slab of the hanging wall in place and when he knocked the reef out, this lump of rock dropped on him, sliding him down the footwall and pinning him to the floor.

Most of the weight was on Yank's chest which made it difficult for him to take a decent breath. Bill O'Donahue was working ten metres below driving off the shaft when he heard the thump and the scream for help and he got up the ten metres in record time. One convulsive heave took the rock off Yank's chest and put the weight on his abdomen and thighs. It must have been pivoted on another rock or Bill never could have moved it. We estimated from its size that it weighed about three-quarters of a ton.

It was a narrow reef and not much ground had been taken with it and, what with the rock and the pinned down man, there wasn't much room to move. Initial rescue attempts were directed towards levering the rock up sufficiently for Yank to be pulled clear. A couple of poles were produced and angled into position but some part of Yank's anatomy was being used as a fulcrum and he gave such a howl of pain and protest that the rescuers had to desist. The only practical course of action was to break the rock into manageable pieces but when this was tried the gads were found to be too blunt, having lost their points on the quartz reef. There was no alternative but to light up the forge and sharpen them. This was always a tricky business, sometimes they would be too soft and their points would turn over at the first blow of the hammer. At other times they would be too brittle and the points would break off. However on this occasion luck was on our side and they stayed sharp until we had broken the rock up on top of Yank and disposed of the pieces down the shaft.

Getting him out of the stope, which opened out directly off the shaft, and up to the surface was not an easy matter, made worse by Yank's continuous groans informing us that he was "all crushed inside". It was midday before we got him up and into an improvised bed in the stores tent. We boiled the billy and sat around drinking tea for quite a while considering what to do next. Jim Donohue, moved by the heart-rending groans from the stores tent, looked Yank over and told us that, "If we didn't do something about Yank's back he'd get fly-blown."

Our medical supplies were rather limited being restricted mainly to Epsom salts which were certainly in keen demand because of our diet. There was however a largish bottle of Fauldings "extra strong" iodine. It smelt like the outpatients' wing of a public hospital and had the consistency of molasses. Armed with a feather from a wild turkey, a protected species but which one of the party had shot in self-defence a few days earlier, Jim advanced on the injured man with the soothing observation that, "This might sting a bit, Yank", a very understated prelude to the application of the iodine with the turkey feather.

Yank's howls of pain dispelled any thoughts that he might have suffered some damage to his chest and lungs and the iodine proved to be an effective fly repellent.

That evening it was decided that Yank would have to be taken to hospital at Meekatharra, one hundred and eighty miles away. We drew straws to see who would stay behind to protect our camp and the remaining stores from a possible visit by a group of aborigines who very occasionally passed through the area. I drew the short straw.

Preparations were made for Yank's comfort on the way to hospital. It was decided to make the trip in the Hupmobile which was thought to be better sprung than the Dodge truck, and some chaff bags were stuffed with spinifex for the injured man to lie on. A start was to be made at first light, which for that locality and time of year would have been 4.00 a.m., and I sat with the injured man while the others slept. About every half hour or so he would say, "I want to pee. Help me up", always to be followed by, "No I can't, I'm all crushed inside, put me down again." I was really quite pleased to see him eventually lifted onto his rather prickly palliasse and listen to the rattle of the impromptu ambulance as it disappeared into the distance.

I then settled down to what I thought would be a three day wait; one day to get into Meekatharra, a day to find out how badly Yank was hurt, and another day to return. On the second day I gave myself a bad fright by getting lost and only finding my way back to camp by good luck or divine providence. The experience rattled me so much that I didn't stir out of camp afterwards. After four days had passed and the party hadn't returned I started to get very edgy and by the end of the week I was in a fever of excitement. The days were long with nothing to do but read and the nights were even longer, lying awake, worrying about the others.

One of the chaps had been a member of a party that had traversed the Canning stock route rehabilitating the wells and had entertained us with stories of how the aborigines had crept up during the night and speared a couple of the party through their tent. Every time the dog barked or there was an unusual noise was a signal to grab the rifle and with ears strained for further unusual noises, await the attack that never came.

By the end of the week of inaction and waiting my nerves were stretched to breaking point. I was lying on my bunk one afternoon and picked up the volume of Kipling's poems that I had taken with me to while away the hours. Reading from where the book opened by chance I read an ominous phrase from *The Explorer* – "Something calling you over the ranges. Lost and calling to you. Go."

Here was confirmation of my worst fears. The party was lost and I was the only person who would know it. What to do? The only place I knew my way to was Peak Hill, a mining ghost town abandoned since the beginning of the century one hundred and twenty miles away. Without any real thought of what I was about to undertake, I rolled up a swag, grabbed a sugar bag and threw a few stores into it, filled a waterbag and, complete with a billy can and rifle, took off for Peak Hill.

I had forgotten about the camp dog, an old fox terrier called Chico. He was so old that all he could provide was canine companionship. It was late afternoon but still stinking hot and we hadn't gone more than a mile down the road when Chico packed up. I refreshed him with a drink of water in the crown of my hat and that kept him going for another half mile. After that I carried him. By nightfall I had covered about eight or nine miles and stopped to boil the billy when I reached a creek from which we occasionally carted water. I decided to leave the dog and walk on, stopping only to boil the billy and eat, a programme which I calculated would put me in Peak Hill in forty-eight hours. I was just preparing to gather up my gear and put the fire out when I picked up the sound of an engine and then raucous voices raised in song. Then the flicker of lights and then the old Hubmobile with the missing party pulled up. "Hi Stump, you old bastard, you out shooting?" Ashamed at my fears, I admitted to being out looking for a 'roo and putting Chico on board, climbed up with him and returned to camp.

The following morning I heard the story of the trip in and a report on Yank's condition. It had been a very slow trip, partly because of the state of the track and the numerous dry, sandy creeks that had to be crossed, mainly by pushing and shoving, but also because of the numerous punctures sustained. Those were the days when under optimum driving conditions a puncture a trip could reasonably be anticipated and an essential part of the tool kit was a pair of tyre levers and a tin of Schrader patches.

By nightfall the party was still a dozen or so miles out of Meekatharra and having used up all of the patches, resorted to stuffing the outer covers with spinifex. The Hup had no lights and Dick Green, an old Hatter who had attached himself to the camp about two months previously, volunteered to walk ahead of the ute and pick out the track. "If I take off my flannel you'll see my white skin." This worked fine until about 10.00 p.m. The party reached the outskirts of Meekatharra when, despite Dick's protests, he was ordered to get back into the ute. "Get back in here you cranky old bastard. We're not driving into town with you leading us in like Stevenson's Puffing Billy." Dick was always one for a quiet life and dutifully climbed into the front seat. In those days the windscreens were hinged to improve either the vision or the ventilation. The windscreen in the Hup was hinged about the middle and when it was open half of the windscreen projected inside the cab. It would have been better if old Dick had been left to guide the party in over the last few chains. The driver put on a spurt and hit a telegraph pole in the middle of the street with sufficient force to throw all the occupants off the tray, including Yank, into a heap in the middle of the road, and surge poor Dick onto the edge of the windscreen. Yank's groans reached such a pitch that his attendants called on Dick with great urgency to "Get some brandy, quick." This little event had occurred just outside the blacksmith's shop which was next door to the Royal

Mail Hotel. Dick wasted no time. Holding his chin in his hand, he rushed into the pub shouting "Brandy, Brandy". Seeing this bearded old gent, naked from the waist up, covered with blood which was overflowing from the cupped hand supporting his chin, the barman wasted no time in pouring out a liberal measure which Dick immediately knocked off and then demanded, "Now another for the bloke outside." Eventually Yank was delivered to hospital. Dick's chin was sewn up and the licensing laws flouted whilst the party recovered from its ordeal.

Medical examination revealed that Yank had sustained some broken ribs and an unspecified injury to his kidneys that kept him in hospital for a further couple of weeks, The attraction of Meekatharra, which boasted four hotels, proved too much for the rest of the party and hence the delay in returning.

The incident was one of those events which have unsuspected sequels. Yank was a personable character who made an impression on the Matron who in turn was friendly with the Caddy family. One of the Caddy family was the mine manager of the Ingleston Consols Mine. Another was the engineer. Yank told his erstwhile lady friend about my accomplishments as an entertainer who in turn passed the gossip on in conversation to the Caddys. This was to serve me in good stead a few months later.

After six months, about twenty tons of ore had been produced, and carting to Peak Hill started. It was the rule that where ore was being carted in from some distance it had to be stockpiled in close proximity to the battery until the total tonnage had been accumulated, following which it was carted into the battery by horse and dray to avoid discontinuity in feeding the battery during the crushing. The old Dodge truck was loaded, overloaded would be more accurate, for the first delivery. It was a nominal 2¹/₂ ton capacity and we loaded it up to 4¹/₂ tons. Wishing the driver well, we sent him off to Peak Hill with his first load. It took him several days to return. The dry, sandy creeks were a formidable barrier and called for a high degree of driving skill and dogged perseverance to get through at all. The chap we chose for the job wanted the job to rotate, but no-one else was game to take it on and we persuaded him to continue. He arrived at Peak Hill with his second load to find that all the old fossickers in the town had thoroughly picked the heap over for the high grade pieces showing gold. This was reasonably easy to do, as most of the gold formed on the footwall contact of the quartz reef and the schist. So here was a quandary. Should he return with the load and pick up another man from the camp to stand guard over the ore, or should he unload it for the locals to pick the high grade out. He chose the second course, an action we endorsed. He said there was no way that he was going to cart the one load three times over that track.

We fell short of the twenty-five tons we had aimed for, managing only four loads of four tons, or sixteen tons in total. About half of the gold was recovered in metallic form, by amalgamation with mercury on copper plates, after crushing by the stamp battery. The other half would only be recovered after cyanidation in leaching vats and for which, some time later, we would receive seventy-five per cent of the gold content. Our expectations then reduced from sixty-four ounces (£256) to thirty-two ounces (£128).

Colin Blakeway and I took the truck with the last load of ore. We got about half way when the truck broke down crossing one of the many steep side creeks that characterised the track. The truck had groaned its way over the roughly corduroyed sandy bottom and was struggling up the bank when there was a sharp report and the truck rolled back onto the creek bottom. A quick inspection revealed that a pin in the front universal joint of the tail shaft had snapped. This was very serious as no other vehicles used the track and it was a long walk in either direction to get help. I rummaged around in the tool box and found a cold chisel of roughly the same diameter as the pin and luckily a hacksaw and a file, both very worn; worn-out would be a more correct description of their condition. With great effort we sawed off a measured length of the cold chisel, a job which took quite some time. I recommend that anyone wanting to saw a piece out of a cold chisel should make sure they have a vice and a sharp saw blade. This accomplished we lit a fire and boiled the billy. We needed refreshment after that episode but worse was yet to come. It was getting dark and by the light of the fire we began the job of filing the pin to fit, holding it in a well worn pair of pliers. It took about four hours to file it down to size. Our anxiety not to overdo it and make the pin too small made it necessary to crawl under the truck dozens of times to try the fit, crawl out, file a bit more off, crawl under again before we got the pin fitted and burred over to hold it in place. That cold chisel must have been made from a very good piece of steel for when we left Nabberu and went our various ways, although we had obtained another pin, the hand-made pin was still holding up in the truck.

The crushing was disappointing. About sixteen tons was crushed for a return of sixteen ounces. We rationalised the low yield obtained compared with our expectations by taking into account the gold that had been lost from the two loads that the locals had picked over, and there was also the gold contained in the residues, but it didn't alter the fact that the gross return was around £70 which had to be split between seven.

After Yank returned to the mine, what with the poor return from the crushing and the trauma of the accident, the party didn't have the same interest in it as before. We were all novices at mining and the stope looked very untidy with a couple or more lumps lying up against the greasy head that had caused the original fall and giving every impression that they would follow suit. One by one, except van Wayland the Yank, the chaps decided they would move back to Perth where they all had contacts of one sort or another. The Yank went to Sandstone where he knew the barmaid. Not having similar attractions, I decided to try for a job in Meekatharra or Wiluna. It was a pretty optimistic decision as jobs were pretty scarce.

Chapter 4

Yank told me I should call at the hospital and in fact took me up to meet the Matron who in turn arranged for me to meet George Caddy, the mine engineer. It must have been my lucky day because the Slimey, the chap looking after the slimes dam, had just chucked his job in. The slimes dam would be the lowliest job on the mine, more lowly even than the sanitary man. The job consisted of building up a wall of slime around the perimeter of the area in which the mill tailings were being pumped to contain the tailings and conserve as much of the water content as possible for recirculation back to the plant to treat further ore. It was hard work and also uncomfortable. It was necessary to work in bare feet with trousers rolled up above the knees. In the summer time the slimes bubbled up scalding hot between the toes and in the early winter mornings, when there were occasional ground frosts, it was painfully cold. However, a job is a job and it was worth ten shillings and a penny a shift on a five and a half day week of forty-eight hours.

The dam had been manned for years by an old Irishman, one Barney Duff, and when he retired and left for a holiday in Ireland his job was taken by a succession of chaps who couldn't stick it out after a week or so, or after a break had occurred in the dam wall through their carelessness and the tailings return water had been lost, a sad blow to the mill boss as water was a scarce commodity. I could not afford to lose that job and worked at it with great care and got the dam into good condition within a couple of months.

Despite the adverse effect of the Depression, an occasional vacancy would occur either in the powerhouse or the stamp battery where the jobs were more interesting and also were paid more. It was my wish to get transferred to one of these but it was a vain hope as I was doing so well on the slimes dam that the boss had no intention of moving me.

The Consols was only a small mine and much of the control was in the hands of an old gentleman named Tom Peart. Tom was the general factotum, sampler, assayer, surveyor, draughtsman and underground manager. By some now forgotten circumstance I became friendly with Tom. I had come by a book on fire assaying and I asked him if I could help him in my spare time. He was pleased to accept and after work I would go to his assay office and help him by grinding the samples down to size, then weighing them and adding the fluxes, ready for firing. It wasn't long before he trusted me with the total operation, a quite reasonable thing to do as the ore did not vary in composition and required only a standard flux. This proved to be a very rewarding association. To get me away from the town, where my lodgings were in close proximity to the hotel, he suggested that I should set up camp in the survey office, an offer which I was very pleased to accept, although shortly afterwards Tom had reason to doubt the wisdom of the move.

It was on the occasion of a National Census that I returned to the survey office close to midnight after an evening with the boys and prepared to complete the census form before retiring. On the large drawing table in the survey office was a beautiful composite plan of the mine workings which had been there from the commencement of the mine and upon which was drawn at each month's end the advances made in the mine headings at each level that had been driven. Forty years of careful work had gone into its preparation and it exhibited beautiful draughtsmanship and calligraphy to which many men had contributed over the years. It was the pride of Tom's heart.

I laid the census papers out on one end of the table and, picking up by the stopper, a bottle of Higgins bright blue ink with which I intended to mark the forms, I moved it across the table. When my hand reached the centre of the plan the bottle became detached from the cork and fell. On impact a large ring of Mr Higgins' product rose from the neck of the bottle like a smoke ring and descended upon and around the bottle which rolled over on its side emptying its contents and tracing in bright blue the path of the bottle as it rolled. I was paralysed with horror as I watched and great damage was done before I reached convulsively for the bottle. My efforts to swab up the mess completed the damage.

When Tom recovered from the initial shock and had eased his feelings, he was very good. I was to redraw the plan. In this way my interest in the technical aspects of the mining industry was aroused and in the back of my mind the thought formed that I could go to Kalgoorlie and develop these new-found interests.

I think with affection of Tom Peart. He apparently saw some latent potential in me and helped me to develop it.

When further opportunities arose for management to transfer me into other jobs with scope for advancement, and I was passed over, I took the only sensible course and left the job to seek work elsewhere. I thought that if I went to Kalgoorlie with the intention of enrolling at the School of Mines I would be better able to get a job in the industry as a miner than as an unskilled labourer. The Triton Gold Mine at Reedy was being developed for production by the Western Mining Corporation and I decided I should try my luck there. Although the Depression was still running its course and jobs for the unskilled were very scarce, there was a shortage of skilled miners to be overcome and Triton management were conducting a school to train development miners. When I arrived at the mine about forty men were at the shaft head each day looking for work. I realised that I wouldn't stand much chance in that competition and taking courage in both hands asked to see the manager. I don't know why he agreed to see me but he did and I stepped straight in by stating that I wished to attend the School of Mines in Kalgoorlie and considered it necessary to learn to be a machine miner before I went there to enhance my chances of getting employment.

Alex McLeod, the manager, was a dour chap, given to long, penetrating stares and disconcerting silences, calculated perhaps to shake the nerve of the interviewee, but I survived the ordeal and he sent for the shaft foreman and told him to give me a job. The initial job was raising by hand via a windlass the broken rock from a winze being sunk on contract by a couple of chaps who wanted the rock out of the winze bottom in the shortest possible time. I had sufficient practise at this to be able to put on an impressive show and, as the winze was being sunk not far from and in full view of the manager's office, my efforts were not wasted. A week later I was told to report to the underground manager. If I thought that I was going to be moved directly into the School for Miners I was quickly disillusioned. I was given a shovelling contract with another chap to move the spoil away from two contract miners. This in itself was good because they were very skilled miners from whom I learned a lot by observation.

The level drives and crosscuts at the Triton mine were driven seven feet by six feet and on the average a five foot advance was achieved per firing. This generated about twenty or so tons of rock which had to be shovelled up into a one ton capacity truck and pushed out to the shaft for the platman to load into the cage for hauling to the surface. The contract rate for this was sixteen shillings a foot and, if you worked like a beaver, it was possible to get rid of it by crib time. It was very important to do this so that the miners could then bore out the lower half of the face and fire at the end of the shift.

The normal procedure when commencing the shift was

immediately to take the venturi bag into the end of the drive to blow out the blasting fumes and then connect the water hose and damp down the heap of broken rock to lay the dust and absorb the fumes it contained. This was of paramount importance because, if neglected, the fumes were slowly released as the heap was removed and would generate a very severe headache in those working on the broken rock. It would take about half an hour to do this and clean up the fly rock before tackling the heap. This meant that to bog the end out by crib time about six tons an hour had to be moved, say a ton every ten minutes. This was about the time it took for the other member of the two man team to push the loaded truck out to the plat and return with the empty. It was hot work and, as I did most of the shovelling, I was always in a lather of perspiration and my clothes sopping wet.

When the first pay day came I was working afternoon shift and collected my pay shortly before I was due to start work and I was in a quandary. If I took my pay with me the notes would get so wet they would be reduced to pulp. I really had no safe place in which to leave them. I was living in a corrugated iron shed about nine feet square, the door of which was fitted with a very impressive padlock and I told myself I should have sufficient confidence to leave my pay in a spare pair of trousers which were hanging behind the door until I returned. When I got back at midnight my pay was gone. An enterprising rogue had lifted the door off its hinges, swung it around on the padlock and helped himself to my pay other than a few pennies which, adding insult to injury, he left behind to taunt me. By the time I worked another two weeks and paid my bill for a month's meals at the boarding house I was back to square one.

Other people had been dealt with similarly and it was generally thought that the identity of the culprits was known. They also worked on the mine and were heavy players at the two-up game. There is good in everyone and if the belief in their identity was correct this was displayed in dramatic fashion a few weeks later. They had drilled out the drive face in which they were working and one of them went out to the shaft for something or other while his mate charged up the face with explosive preparatory to firing it. The customary practice those days was to light the fuses individually and as about twenty holes would have been drilled the face was always heavy with smoke by the time the last fuse was lit. Deciding to return and either meet his mate coming out or help him with lighting up, he found that his mate's retreat was in serious hazard from a crosscut in which the fuses had been lit and which his mate had to pass to get to a safe distance from his own blast.

A steam powered diamond drill operating near Reedy's, W.A., at what later became Triton Gold Mines N.L. c 1934.

Triton Gold Mines exploration crew, British Bedford truck and steam boiler en route.

Without hesitation he rushed in to get his mate out who in turn had just lit up oblivious of the hazard abreast of his retreat. There followed a frantic rush down the drive, past the crosscut and around a protecting corner when both faces exploded. This experience and the probable awareness that there was general suspicion that they were responsible for the many thefts prompted them to leave.

It was not long after this that the mechanical bogger was introduced and nowadays a miner asked to use a shovel for any other purpose than a bit of cleaning up where a mechanical device could not be introduced would recoil in horror at the thought.

After the war, when underground labour of all sorts was undersupplied, the Chamber of Mines of Western Australia recruited large numbers of Europeans – Italians, Dutch, Yugoslavs, Balts and others – to fill the gap. They were tough, hard-working people who stayed with the industry and were its backbone for the next twenty-five to thirty years. There were of course a few Australians among those recruited. One such was serving his orientation week on the three thousand eight hundred foot level of the Lake View and Star's Chaffers Shaft. On his first day, whilst leaning on his shovel wearing a pained expression, he came under observation of the underground manager. "What's the matter, lad?" he asked. "You feeling crook or something?" The young fellow confessed to feeling off-colour, a condition that the underground manager attributed not to a feeling of great repugnance for the shovel, but for a genuine malaise. "You probably need a bit of fresh air. Go up top for a quarter hour or so. That'll fix you," he said as he rang for the cage. It arrived and he rang the signal to hoist him to the surface. Just as the cage was disappearing he had a terrible thought. "You are coming back, aren't you?" The answer came back just as the cage disappeared from view, "Not unless the rope breaks."

For me in 1934 that contract on the shovel was a welcome break. I didn't have it for long as after two or three weeks I was moved into the School for Miners.

The school was an initiative of G. Lindesay Clark, later to become Sir Lindesay Clark, Chairman of Western Mining Corporation, Alcoa of Australia and Broken Hill South, but at that time Managing Director of WMC.

The school was under the direction of a shift boss named Walker, who had been recruited in Broken Hill. He had three trainees at a time and the course duration was three weeks. Commencing with a week on day shift, it progressed through night and afternoon shift at the end of which the trainees were assigned a development end on contract and turned loose. During the training period the hopeful miners-to-be were instructed in the different drilling patterns that could be adopted in a variety of rock conditions, in the care and maintenance of equipment and in the use of explosives. It took some time after being discharged from the school to develop the skills necessary to ensure that a reward could be earned commensurate with the effort expended. I recall that on my initial two week contract I earned only half the minimum wage at the expense of consuming one half ton of gelignite and three-quarters of a mile of fuse. The number of detonators used has escaped my memory but I recall with some embarrassment that many faces had to be reblasted. However with such practise it was inevitable that there would be improvement and it was not long before I achieved results that I think vindicated the decision to include me as a trainee.

It was very hard work. Once the development cycle was in full swing, it was customary to rig a bar horizontally across the face just above the level of the heap of broken dirt resulting from the previous blast. The bar was four inches in diameter of heavy duty construction with a screw foot with which it was held firmly in place by tightening the screwed foot against wooden blocks between the bar and either wall. Two three-inch diameter arms three feet long were clipped onto the bar with heavy bolts to support the machines which were similarly clipped to the bars. The clips could be loosened to allow the arms or the machines to be moved to new positions either along the bar or up and down the arms.

The machines, formidable pieces of equipment weighing one hundred and sixty pounds, were supplied with compressed air through a heavy duty two inch reinforced rubber pipe and a one half inch water pipe. During the course of drilling a face, the assembly had to be manually raised to the limit of the arm so that the top holes could be drilled a bit more than two metres above the floor. Once this feat of strength had been performed the machines were lowered progressively and holes drilled in a predetermined pattern down to the bottom holes, the lifters. The cycle was usually completed in one shift, the last hour or so being taken up with dismantling the gear and carrying it back to a safe spot where it would not be damaged by fly rock from the next firing, laying the sollar boards upon which the bulk of the rock would fall after firing to give a smooth surface to make shovelling easier, blowing the holes out, charging up and finally lighting up.

It was heavy work for which no comparable effort is required today. Manual drilling where it is still employed now uses light, fast-revving machines supported pneumatically on light air legs, otherwise the drilling equipment is crawler mounted and self-propelling. My own shift was always an hour or so longer than anyone else as I managed to get myself so much more begrimed with grease, pneumatic oil and dirt that I spent anything up to one and a half hours each shift under the shower getting clean.

I stayed on that job for about six months. It was invaluable experience. When in a position of authority and dealing with industrial disputes or claims for more pay, it is of great importance to know the type of work and the associated physical and nervous stresses under which your people work. In the process a good understanding of the workman's psychology is gained. Modern technology, equipment and management techniques can be bought. The loyalty of a workforce which is of equal importance can only be earned. The graduate engineer who comes straight out of university into industry without the benefit of industrial experience is under a handicap in relation to those who have come up through the ranks. This is evidenced in the industrial record for Kalgoorlie which has been virtually strike free for its entire life. Its managers have grown up in the industry and obtained their professional qualifications at the Kalgoorlie-based Western Australian School of Mines. Good industrial relations are built on mutual respect.

After spending six months as a contract development miner I suffered a minor accident to my hand which kept me from work for a couple of weeks. I returned to Meekatharra, where most of my friends were, and was offered a job as field foreman with an exploration team. It was not a very important position. There were two diamond drills operating and another small party rehabilitating a shaft which had been sunk probably forty years previously and abandoned within a few years when the gold cut out. I reported to an elderly man named Martin, known familiarly as "Plugger". In those days everyone named Martin was called Plugger, a tribute I believe to a professional cyclist who earned the sobriquet in recognition of his dogged persistence. It was a welcome change supervising others instead of being supervised and it rekindled the flagging desire to get some qualifications at the Western Australian School of Mines.

I turned my thoughts to returning to England to see my mother as a desirable prelude to getting back to Kalgoorlie in time to enrol for the 1936 school year commencing in February.

My main problem was financial. I had been getting around with a group of people a couple of tiers above me in earning capacity and although I had been in continuous employment I spent what I earned as fast as I earned it. I was hard put to rustle up the fare on a one-class boat from Fremantle to Southampton.

The voyage duration was thirty-one days and the fare £35 Australian. What a measure that is of the extent that inflation has degraded the value of our currency.

At that time, in my position as exploration foreman, I was earning £40 per month. By dint of rigorous self control over my social habits I managed to pay my fare to Perth and my fare to England on the *Jervis Bay* with a little more than £20 left over, which I transferred to the London branch of the Bank of New South Wales.

It was a happy voyage. There was a full complement of passengers, predominantly young people visiting the UK and Europe on holiday and intent on having a good time. I was probably the only passenger who wasn't provided with spending money for the voyage or indeed for the holiday and return trip, a condition upon which I look back with a strong sense of humiliation.

The long nights at Beacon playing bridge whilst waiting for a job, any sort of job, to turn up stood me in good stead in this situation and I managed to win sufficient money playing bridge on the boat to keep me in the bare necessities of life. I smoked in those days and I also had strong yearnings for the company of young attractive women.

My lack of money was a real test of the depth of feeling held by the one or two girls with whom I spent my time. It made me recall the practice of a young man with whom I became friendly during my early days in Australia. He said that whenever he acquired a new girlfriend, on taking her out for the first two or three occasions he would take her to the cheapest seats in the cinema and buy her a bag of peanuts. He figured that if she stood two or three weeks of this behaviour she must like him for himself alone and not for a good time at his expense.

Shipboard romances are wonderful experiences often with painful emotional terminations. My involvement with a New Zealand girl was no exception. It was brought to a painful end when my mother swooped on the boat train at Waterloo station and whisked me away.

Before setting out on this return visit to the land of my birth and the familiar sights and people associated with my school days, I had looked forward with pleasure to renewing some of the friendships I had made at school. What a disappointment I had. Perhaps my recollection of them as bright imaginative chaps had become enhanced with the passing years for I found them to be very dull people. Perhaps I was put out because they exhibited no interest in my very diverse activities but insisted on talking about their own very uninteresting lives. With all the egotism of youth

I thought gratefully of the wisdom of my parents in lifting me out of the environment in which in all probability I would have grown up like them.

I had enjoyed my time at Meekatharra very much. It was an unusual town, totally devoid of any social stratification but totally polarised as to religion – Roman Catholics versus Masons, It was a very small community of not more than seven or eight hundred people all told. The main source of employment was the Ingleston Consols Mine which employed about one hundred and twenty people. A normal ratio of dependants to employees would have been, in those days, about four to one which would account for about six hundred bodies. And then there were the normal services required in which the hotels were somewhat oversupplied. To satisfy the thirst of this small population, plus passing traffic and the prospectors working the surrounding country who would come in on occasion to slake their very considerable thirst, there were four hotels. There were tradesmen there that are rarely seen in small country communities these days – a blacksmith/farrier, a saddler and a tailor who, with his assistant, would sit cross-legged on a table when hand sewing parts of the garment that could not readily be got at with the sewing machine. A good made-to-measure suit – which really fitted – and was made from good woollen cloth, cost £10. A suit of that quality was a must for social occasions such as a concert or a ball! It didn't matter what sort of clothes a person wore during the working day but everyone smartened up for evening functions, ably assisted in the choice and quality of their attire by the local haberdasher and men's outfitter John A. Bell who boldly advertised his wares as being "For the man who cares."

Alexander Turner, a young teller at the bank, was very talented in the production of concerts and drama, and I performed in several of his productions. In the years following the war his name became a household word from the radio plays that he presented for the ABC every Sunday night.

When the need arises, people are great improvisers. The summers were very hot, nearly every night the temperatures would rise to a level at which in Perth would have people flocking to the beaches. There were a number of sandy-bottomed creeks within a reasonable distance of the town which the locals had named after city beaches – Cottesloe, Scarborough, etc. They were of course bone dry but they were good places to resort to for an evening barbecue. It always seemed to be a bit cooler out there and, even if it wasn't, it took your mind off the heat.

Apart from there being no ocean for swimming, boating, fishing or the opportunities to admire girls' figures reclining seductively on the beach, Meekatharra had most of the other forms of recreation – golf, hard court tennis, cricket, football and the minor amusements of billiards, darts and hookey, which are games usually played in the front bar of pubs. In golf the complete absence of grass was compensated for by the tremendous run of the ball which would often give even a novice a drive of three hundred yards, to be followed by a careful dressing of the sand green with the scraper, an essential part of every golfer's kit, assuring a guided path to the hole. These were conditions which gave an average player the status of a low handicap.

Casting my mind back to the cricket matches that were played every Sunday during the summer, I recall my amazement at the conditions under which the game was played; a concrete pitch covered with coir matting in the middle of an oval of bare packed earth or gravel from which the heat could be seen rising in shimmering waves. Football was played in season on the same ground and if any one present today thinks today's footballers are a hardy race, they should have seen the country games of seventy years ago when players would be brought down on rock hard gravel or concrete.

One very popular activity for which I have difficulty in finding an appropriate name, for it could be described as a community duty, a sport, a hobby or athletics, was membership of the Volunteer Fire Brigade. This wonderful movement established itself in every country town to deal with sudden outbreaks of domestic house fires. It calls for a special sort of dedication as its members are on call every hour of the day or night throughout the year. This service, which was essential in small towns unable to maintain a paid service and as often as not without a reticulated pressurised water supply, was also a sport which reached an annual culmination in the Easter Trials. Practically every country town, and some metropolitan suburbs, fielded a team. Every town had its training tower and on Sunday mornings the crews would rehearse with great enthusiasm the drill of racing down a measured track tugging a two wheeled cart on which a hose, reel, hydrants and nozzles were mounted, progressively connecting them to the water hydrant and carrying hose and nozzle up a thirty foot ladder, usually taking two rungs at a time, to aim an imaginary stream of water at a target. Where there was a copious water supply, sufficient to justify the use of water during training, the trick was to run and climb fast enough to beat the stream of water to the top, thereby avoiding carrying the weight of the water as well as the hose.

This is one sport (I think I can call it that) that has persisted until the present and the Easter Trials still attract a large number of teams and many admiring spectators.

My circle of friends consisted of the doctor, the Roman Catholic priest, Father Prendergast, who I always thought of but never addressed him as "Old Prendy", one of the local publicans, a not surprising inclusion as we mainly met in the corner of his front bar, the local bank manager, and a couple of local "gentlemen prospectors". This little group generally met in the evenings after work and on Saturday afternoons. The time was usually spent drinking beer, playing cards to see whose unlucky fate it was to pay for the drinks, and generally discussing a variety of topics. I suspect it was the Holy Father who inspired and kept alive the conversation as he very rarely lost a game and only very infrequently had to pay for the drinks. It is difficult to maintain a sound line of argument and exercise a high degree of skill at the same time but he was very accomplished in both conversation and card playing. He really was a most unusual person.

There was a legend that he had been in the intelligence service (for that read spy) in the 1914-18 war. No one ever verified this but I was prepared to believe the legend as he had several languages and was a very cool customer. The fact that he was quietly tucked away in Meekatharra suggested that he had fallen from grace with those in the higher echelons of the Church. From my recollection of him he was about seventy at the time of which I write. He had no motor vehicle of his own but hitched rides between home base and outlying parishes. On returning one day from a visit to Wiluna with a commercial traveller many years his junior, the car broke down about twenty miles out of Meekatharra. In those days it was not a frequently used road and, rather than wait for a passing vehicle that could get mechanical help, Prendy elected to walk the intervening miles. It was a hot day and after settling the driver under a shady tree with instructions not to leave, the old boy took off arriving in Meekatharra late in the evening, having been given a lift over the last few miles and, after a couple of refreshers insisted on returning with the mechanic to show him where he had left the driver.

On one such occasion in the bar the conversation turned to the efficacy of prayer which as may be imagined the Holy Father was strongly defending on grounds of personal experience. His case was centred on an experience when he was travelling to Sandstone, a small mining town about one hundred and ninety-five kilometres south of Meekatharra, passing en route a mine at a place named Errol's which was owned and worked by an Irish family. This was a usual stop-over for him as they were of the faith and very good supporters of the Church. After lunch, when it was time to resume the journey, the head of the family asked him if he would bless the mine. "Shure it has been very good to both of us," he tactfully pointed out, "and now we be sinking the shaft and if your Holiness could bless it?" Here he paused hopefully and Prendergast gladly agreed. As he told it, while the party was walking up the hill to the shaft-head he searched his missal for a prayer suitable to the occasion. "I couldn't find a blessing for a mine, but there was one for a well and I thought, hm, it's not a mine but they are both holes in the ground and this one will do. They had the shaft bottom bored out and charged up and shortly after I left they fired and encountered a strong flow of water. It proved to be uncontrollable and they never got another ounce of gold out of the mine."

Like a lot of Prendergast's stories, they were told with such sincerity and attention to detail that you never knew whether to believe them or not. Certainly at that time the mine was abandoned.

Towards the end of my stay in Meekatharra I had an experience with the Holy Father which demonstrated his skill at extricating himself from unpleasant circumstances. Having been working all of one Saturday afternoon it was late, about sixish, when I walked into the front bar of the hotel to find Prendy propped up in his usual corner and I regret to say rather the worse for wear. He had always seemed to be impervious to alcohol but on this occasion he had obviously over-indulged. On getting me in focus, and recognising a friend to call on in his hour of need, he asked if I would drive him back to the Presbytery. Not suspecting an ulterior motive I agreed and assisted him into the front seat of my utility. As we were approaching the Commercial Hotel he asked me to stop and buy two bottles of ginger beer. It seemed an odd request but people often have strange remedies for over-imbibing. But I was wrong in my assumption that it was for his personal use for, as he was nearing the Presbytery, he revealed tactics which I suspect he had used on previous occasions. "You take the lead, my boy," he said, "And I will follow. The Mother Superior will open the door and you are to immediately present the bottles of ginger beer to her. I will then announce you as my dinner guest and with minimum delay we will proceed to my quarters."

The expression on the Mother Superior's face when she observed old Prendy's condition told me he was in for trouble and that I should follow him as quickly as possible. He led me into the dining room where the table was set for one and told a very dismayed Sister that I was staying for dinner. The portion that had been put aside for him was very meagre and when it was divided into two parts it was almost microscopic. Perhaps he was still thirsty or more likely he thought he should recompense me with a drink and fortify me for the ordeal that awaited me on my departure for he bade the Sister to bring a bottle of wine. "Shure we don't have a bottle in the house," said the Sister. "Away and bring a bottle of altar wine," said Prendy. When it arrived it proved to be sweet sherry. We had a couple of glasses each after which he probably thought that I was sufficiently fortified for he said, "You must go now, my boy. Find your own way out and look out for the Mother Superior. Good luck." His good luck wishes were to no avail. The Holy Mother was waiting for me, barring my way to the door, until she had told me what a scoundrel I was, playing on an old man's weakness and then having the effrontery to bring him home and eat with him and partake of wine reserved for holy use. She had a fine command of language and exceptionally good lungs. When she had finished she opened the door and with an imperious gesture indicated that I should get out, quickly, leaving no opportunity for me to explain that my part in the Father's fall from grace had been solely to play the good Samaritan.

I saw much less of the Anglican priest, the Reverend Picton-Warlow, who was not resident in Meekatharra and had an enormous parish to service. The Anglican Church was not strong in Meekatharra, the population being either Roman Catholic or Methodist, which seemed to be the popular church for the Masons. My chap used to visit about once every two weeks or so and I was often the only worshipper present at his services. He was rather deaf and not being able to hear my responses and with his nose buried deep in his prayer book would launch off into his bit before I had finished mine so that at some of the time it would have sounded as though we were performing a duet. He had such poor support from his community that I would attend as often as I could.

The Methodists were the other strong faction which was composed predominantly of Masons. Mr Dick Caddy, the manager and part owner of the Ingleston Consols mine, around which the economy of Meekatharra revolved, was a very dedicated Mason. Being a public spirited man he was also chairman of the Roads Board upon which there was a strong representation of RCs. At the time of which I write illumination underground was by candles and the daily issue was drawn by the underground workers who filed past the candle house to receive their issue which varied according to the type of work upon which the recipient was engaged. As the man issuing the candles was a staunch Roman Catholic, the number frequently varied with the applicant's religious allegiance, a fact that was slyly inferred at a meeting of the Roads Board during a deliberation on the choice of street lighting. The question whether it should be by acetylene gas or electricity had been vigorously debated for several hours without a decision being reached. During a pause for breath during the discussion, one of the members, an Irishman, blandly suggested that candles might be a suitable alternative. The reaction of the chairman was immediate and explosive. As mine manager he "knew all about the cost of candles, didn't the idiot who made the proposal know that candles were the most expensive form of lighting there was," etc, etc. When he eventually subsided from lack of breath, the Irishman slyly remarked, "Shure you surprise me but I'll take your word for it. I just know it doesn't cost anything to light our Church and we use candles all the time!"

That was one of the more humorous episodes in the antagonism between the two groups which would find expression in such comments as, "Did you meet that new chap that got a start at the mine today. He's a Mick alright. You can tell by their eyes, you can never trust them. Yes, he's an RC alright." Or it might be, "That new chap, he's one of them, he's a Mason. Look at his legs, he's bandy. That comes from riding the goat! Look out for that chap."

That is reminiscent of an incident in Norseman several years later when a man in a very happy state of advanced inebriation, walking down the main street one Saturday evening, was lured to the main doors of a hall by sounds of merriment within. The local branch of the Royal and Ancient Order of Buffalos was celebrating after concluding their official business. The drunk hammered on the door and when an eye appeared at the peep hole to investigate he demanded to be let in. "Are you a Buffalo?" inquired the eye. "Don't know about a buffalo," the drunk replied, "but I've been working like a horse all the week."

During the later part of my stay in Meekatharra I had become the proud owner of one ninety-sixth part of a local syndicate formed to dewater the Ingleston Alberts mine on Paddy's Flat and investigate the chances of bringing it back into production. The main shareholders were the Rinaldi family and a local storekeeper, Bert Gerick. Bert, who was the syndicate manager, was also the winder driver. His wife kept the shop while he was working on the mine. I think the bottom level was about six hundred feet below the shaft collar and the mine had been dewatered and kept dry by bailing, using tanks specially designed for the purpose. After the bottom level had been exposed, the mine was effectively kept dry with one tank only, the other rope on the two drum winder being connected to a man cage for raising or lowering men, supplies and development ore or muck. At the time I was the sampler and assayer, a job which involved going underground at 7.00 a.m., while Bert was bailing the water that had accumulated since the previous day, and sampling the development end before the miners commenced work at 8.00 a.m. The shaft was always very wet from spill from bailing and on the first morning I decided to keep myself dry by wearing oil skins. Bert was quick to react. "What have you got those on for?" he asked. "What do you think, to keep me dry", I replied. "They won't do you much good", said Bert. I failed to catch the glint in his eye. When the cage reached the bottom level with me in it, it kept on lowering until I was up to my neck in water in the shaft sump. It was a very spooky feeling not knowing whether it was going to stop before I was totally underwater and while I was trying to raise the bonnet of the cage before I was totally immersed he raised the cage to let me out at the level plat. It made his day and kept him laughing every subsequent morning when I went underground minus the oilers and when he stopped at the shaft brace momentarily to let me off before lowering the bailing tank on the other drum into the sump.

Among the equipment on the Alberts mining lease, held under machinery lease by Claude de Bernales, was an air receiver and a two drum hoist which the syndicate needed and sent Bert to Perth to negotiate its purchase. After a one thousand kilometre rail trip to the city over a very poorly ballasted track Bert would not have been in a genial mood. The train arrived at 9.00 a.m. and he went straight round to meet Claude who was as usual immaculately dressed and quite effusive in his greetings as he tendered Bert a Havana cigar from a heavy silver box. Bert didn't fall for that sort of treatment and was not intimidated by the spacious office panelled from floor to ceiling with beautiful West Australian jarrah. Claude, sensing the atmosphere, quickly came to the point. "Now what can I do for you Mr Gerick." Bert equally quickly stated his case. Claude summoned the accountant to request the price at which the equipment was carried in the books. The accountant returned and deferentially handed Claude a piece of paper, face down, upon which presumably the price was written. Bert said Claude turned up one corner and scrutinised it like a man "squinting at the hole card he held in a stud poker game".

"Ah Mr Gerick, our price for the two items is £1,500, they would of course cost more if purchased as separate items."

"No go", said Bert, "too dear".

"And how much are you prepared to pay Mr Gerick?"

"Eight hundred", said Bert, "and not a penny more."

"That is a very large reduction below their real value Mr Gerick which I am afraid I could not accept. However there is another way around this; I would be prepared to exchange the machinery for equity in your syndicate and to undertake to float a public company which would be profitable to both of us."

Bert exploded. "Not on your bloody life!"

"Why ever not Mr Gerick?" asked Claude in some amazement at the force of Bert's reply.

The answer was swift in coming. "You're too much of a crook for us!"

It took about two seconds for the rebuke to sink in and then Claude threw himself back in his chair and laughed his head off. When he had wiped his eyes and recovered his composure he said, "Very well Mr Gerick, yours for £800".

Bert returned well pleased. A couple of years later, and after the expenditure of much more money on the development of the mine, the Board decided to send Bert down to see Claude again and ask him to renew his earlier offer of participation, but without success. "Ah Mr Gerick, two years ago I could have done something for you but I am afraid the time has passed."

Shortly afterwards development was stopped and the plant and equipment sold off. Every now and again I would receive a cheque for a small amount such as £1/4/9 representing one ninety-sixth part of the proceeds of one of these sales.

A few weeks before I sailed for England, home and mother I was engaged by the Bernales group to sample workings on the "Haveluck" lease. This was situated on one end of a porphyry dyke about a kilometre or more in length in which a few shallow holes were evidence of earlier half-hearted interest by the occasional fossicker. The main workings were a large spherical "ballroom" about ten or twelve metres diameter which had stopped short of breaking through to the surface by about two metres.

In these circumstances it was difficult to get a truly representative sample of all the exposed surface but the values I obtained were so uniformly low grade that I was satisfied that the weighted average of 4.5 gms/tonne was fair and reported that figure.

A couple of months or so later, sitting at breakfast with my mother in London and reading the *Telegraph*, I turned to a page fully taken up with extracts of a prospectus for a company to be called Meekatharra Gold Mines NL. The description of the prospectus which was the old "Haveluck" lease was much at variance with what I knew about it. The average grade quoted at very much higher than my sampling results was the most blatant example, but there were other totally misleading statements of ore reserves, broken ore stocks and the inference that the ore persisted for the entire length of the porphyry. I was appalled and, after talking it over with mother, I decided that I should go up to the city and tell the managing director what I knew.

When I presented myself at the London office the managing director wasn't available but I could see the chief mining engineer. He listened to what I told him without showing any emotion. When I had finished he asked, "Is that all?" I thought it was more than enough and he must have thought likewise because he showed me the door and bade me "Good morning" without any further words passing between us. The float went ahead with Claude taking a large slice of the money raised, in cash as vendor's consideration. When I returned to Australia I went directly to Kalgoorlie and have no further first hand knowledge from that point other than that the operation survived only a few months. During the 1980s, with the aid of modern large materials handling equipment, it was profitably mined at the surface and heap leached.

I had many friends in both camps and I was sad to leave Meekatharra but I knew the time had come to make a break. During the three years I had spent there I had got something out of my system. I had worked hard and spent almost everything I had earned wastefully on keeping up with the Jones'. The only thing I proved was that I had a very hard head for strong drink and that I was accounted good company. I had in the process got that nonsense out of my system and had formed a firm resolve that, having an industry and environment that I enjoyed, I was going to make a position for myself in it.

Chapter 5

On returning from England my arrival at Kalgoorlie was very low key having travelled by rail on a third class ticket and with only £2 to my name. The tariff at country hotels in those days was usually £2 a week and fortunately it was paid in arrears. Jobs did not appear to be too plentiful and it was a stroke of luck when I learned that the manager of the Kalgoorlie Ore Treatment Company was a fellow traveller on the *Orama* which unloaded me in Fremantle at the end of November 1935. I considered this to be sufficient justification for me to claim special preference in my job-seeking.

I immediately wrote to him and referred to this unique coincidence pointing out the difference in our circumstances. He had travelled first class to an assured managerial position whereas I was numbered among the third class passengers and had neither money nor a job waiting for me but was most anxious to find any sort of work, so that I could study at the School of Mines and learn how to be a manager also. Well, perhaps not those exact words but along those lines.

It must have been a persuasive letter because I received an early reply telling me to present myself at the manager's office for interview. Not wishing to dispel the favourable image that I believe my letter must have created to have got me to the interview stage, although I expected at best to be given a low grade job of a labouring nature, I attired myself in my best garb, a suit that I had made to measure in London by a firm that was doing a roaring business operating under the name of the Fifty Shilling Tailors. Don't get me wrong, this was no tatty little firm but a large business that operated in a big way throughout England. The Depression still held the UK economy in its grip. Their fifty shilling suit was the standard but there was a premium grade available for seventy-five shillings and I had a suit of this quality made from a very nice light grey gabardine that I had selected from a very large range of patterns. The tailor did an excellent job and it fitted me perfectly. Arrayed in this, I set off with highly polished shoes, a neat tie and a smart velour hat (everyone wore a hat in those days). I caught a tram out to the mine and, in crossing the lease in search of the mine office, asked direction from an elderly gent who I found leaning against a desk in the main mill-room. Not content to just point the way he insisted on accompanying me to the building in question and enquiring on my behalf for the manager who at that moment was out around the lease somewhere. After finding a chair, this very considerate fellow pressed me to sit down while he went in search of the manager whom he brought to me. This took about fifteen minutes and when the mission was satisfactorily accomplished he accepted my thanks for his courtesy with a deferential, "Pleased to be of help, sir."

The interview went off satisfactorily with only one hiccup. It was close to Christmas and the plant was shutting down for two weeks for annual maintenance. I could come back in January and would be employed. This was no help as I had spent the small amount of money with which I had arrived in Kalgoorlie and needed a job immediately. After some very heavy special pleading incorporating the statement that if I couldn't get a job straight away I would have to leave Kalgoorlie to seek employment in another part of the state and my hopes of attending the School of Mines would be dashed, the tears were not exactly coursing down his cheeks in sympathy but he agreed that I should commence work next day and told me to report to the mill foreman at 8.00 a.m.

This I duly did and found to my dismay he was the same old gent, no longer kindly and solicitous of my welfare, who had been of such help to me the day before. His face revealed his chagrin as he looked first at me and then at the memo in his hand and menacingly said in a malevolent voice, "So it's you." I thought it polite as a subordinate at the very bottom of the pecking order to reply with due humility, "Yes, sir."

The first job he allocated to me was cleaning up the spill of fine ore under the conveyor belts to which I applied myself with zeal and finished in an hour and a half. When I reported back to his desk he wasn't there, but when he returned and found me sitting on his stool he was full of indignation.

"Get back under those bloody belts and don't come back here again until you've finished." He was even more indignant when upon inspection he found the job was finished and in good style. He gave me another task with the irritable instruction to "Make this one last." An unpromising start, but in the long run he turned out to be a valuable ally.

Enrolment day at the School of Mines was at the end of February. Enrolments were at a very high level by people of ages ranging from fifteen to fifty-five. The Depression still had large regions of the state in its grip and there were many men in my age

WA School of Mines Laboratory, 1930s.

group with similar ideas to mine from both city and country. These were those who took the course part-time while working full-time on the mines.

The course offered in those days was an Associate Diploma spread over four years of full-time study. The lectures and laboratory work offered during the mornings from Monday through Friday were duplicated in the evening to accommodate those working on rostered shifts. For those students who had not completed the leaving certificate, or were of a mature age and whose earlier studies needed refreshing, there was a preparatory year in Maths, Chemistry, Physics, Geology and Mechanical Drawing. As it was eleven years since I left school in England, I opted for the preparatory subjects which on full-time study took one year to complete. By this time I had a permanent job as a plant operator. Treatment plants have traditionally operated on a year round programme of three eight-hour shifts a day, seven days a week, 365 days a year. In those days it was customary to work by calendar months with one day off per half month so that pays were either for fifteen days or sixteen days excepting of course February. Also there was very little time off between changes of shifts at weekends. There was an eight hour "quick shift" change from afternoon shift to day shift and from day shift to night shift, and another, rather ironically called the long weekend which occurred every third week from 8.00 a.m. on Saturday morning until 4.00 p.m. on Sunday. This was the time to get relief from a very exacting regime of work and study. This took the form of reporting at 1.00 p.m. Saturday at the saloon bar of the Palace Hotel which, for what it is worth, was the most prestigious bar in Kalgoorlie. There a little coterie of friends of like mind and similar habits met and drank steadily for the next eight hours whilst debating with great fervour a range of topics – cultural, philosophical, political, scientific, literary – you name it, we discussed it.

The war-like preparations of Germany, the threat to peace they posed, the likelihood of war and the effect such an event would have on our studies featured largely in our discussions. By 9.00 p.m., the effect of a large quantity of beer taken on empty stomachs would prompt the party to break up, each to find his way to whatever he called home, the lucky ones changing from night shift to afternoon shift consoled by the knowledge that they didn't have to go to work until four o'clock the next afternoon.

I was doing rather well at work, having been moved through a variety of jobs to a point where I could be called upon at short notice to fill in for almost any one who failed to turn up or was sick. It was about the time when froth flotation had just been introduced locally into the treatment process as a means of separating from the gangue and concentrating the sulphide minerals which in Kalgoorlie were pyrite with small amounts of telluride. These minerals contained about sixty per cent of the total gold in the ore, of which about ninety per cent was recovered, the other thirty per cent being individual particles of gold which were recovered by a gravity concentration that hadn't changed very much since Jason stole the Golden Fleece.

Previously, to oxidise the sulphide minerals in the ore and render it amenable to cyanidation, it was customary to roast the total ore at a temperature of about five hundred degrees centigrade in a long narrow reverberatory furnace along which the ore was continuously moved by rotating rabbles from the point of entry to the discharge. This then went through the standard cyanide process of agitating the ore with cyanide solution, separating the gold solution from the now near barren ore by thickening and filtering, clarifying the gold solution for precipitation of the gold on zinc, and finally separating the gold by smelting the zinc/gold precipitate.

For over thirty years timber cut from the region's forests had provided the heat for this process. It was not cheap fuel and the need to roast the total ore had proved so expensive that the treatment cost had caused many mines to cease operation. It also resulted in severe denudation over a radius of one hundred miles of the native eucalyptus woodlands which characterise the Kalgoorlie region.

The flotation process, the use of which recovered a high percentage of the sulphide content of the ore in a high grade concentrate, allowed the demineralised gangue to be discarded. The high sulphur content of the concentrate made it self-oxidising in the furnace eliminating the need for external fuel. The combined cost saving in fuel and the need to cyanide only the small sulphide fraction of the ore brought very large tonnages of ore previously subeconomic back into the category of profitable ore again. It was of great economic importance.

The flotation process had only just been introduced into the treatment plant where I worked and it appeared that no-one knew very much about it. I became aware of this when I was called in to operate the flotation section for a week of night shift. The process as operating was a depressing sight. Instead of a steady flow of bean size concentrate-covered bubbles rising from the flotation cells in a continuous column to the surface to flow over the apron via the launder into the concentrate collection tank, a few large green bubbles, football size, would rise reluctantly to the surface, burst with a small plopping sound like the mud bubbles at Rotorua, and subside to the bottom of the cell.

After attending to this system for an hour or so, I was unable to resist interfering with it. The type and quantity of reagents to be added at a regular rate per minute had been posted in a minute book for the benefit of the operators. These were carefully metered by a calibrated reagent feeder which fed with a high degree of precision into the conditioning head tank. After a couple of nights I had reorganised the reagent additions both by quantity and also by point of addition. I had procured a number of four gallon kerosene tins which I filled with one or another of the various reagents in use from which I could adjust the flow by perforating the bottom edge of the tins and inserting a piece of wire of a size which would give the required amount. These I positioned at various control points in the system.

I got an immediate small improvement and by the end of the second night the system was starting to work the way it should. I could hardly wait to get to work each successive night to recover my apparatus from where I had stashed it, reassemble it and make more variations to further improve it. By the end of my week's night shift I had it working really well.

On the following Monday, shortly after commencing work on the afternoon shift, I was confronted by the manager who questioned me searchingly on the manner in which I had taken the samples at various points along the process by which he monitored the effectiveness of the reagent regime. I had a clear conscience on this as I had taken the samples precisely at the time and in the manner specified. This caused him to enquire how it was then, as the three operators were all working under the same conditions, that the results obtained on my shift, as assessed by the samples taken, were the only ones that approached the level of success for which he was aiming. Then I told him with a high degree of apprehension for my future employment, what I had been doing. My job hung in the balance. I recall he was most outspoken on the subject of employees who went about aborting their superior's work. I had no idea that he had such an extensive vocabulary. When he ceased frothing at the mouth and was able to review the situation objectively, he said, "Now, tell me exactly what you did, what amounts of the various reagents you added and where."

Next morning a plumber arrived to extend an array of small diameter pipes equipped with pet cocks from the reagent feeders to the points of my addition along the flotation cells. It wasn't long after that a new regime of reagent addition was adopted which was progressively modified until consistently good results were obtained.

It was with some satisfaction that during a subsequent discussion between the manager and plant superintendent, in which there was a considerable difference in the views each held, I heard the manager say, “But Brodie-Hall says!”

Until I was promoted to the status of foreman, working only the day shift, it was very heavy going. Night shift was the killer. I never acquired the ability to sleep during the day. To keep up with the full-time syllabus I had to spend a lot of time in private study and had it not been for the good will of the shift bosses who would take over my job at about 2.00 a.m. for an hour while I got a sleep in a quiet corner, I doubt if I would have been able to continue with the full-time course. They were very good-hearted people who, having had a minimum amount of schooling themselves, were pleased to be able to help another person to (as they would have put it) improve himself.

Everything has its price and at the beginning of the second term of Year Three when some of the subjects were rather meaty, I had indications that I was not well.

It is a disquieting experience to suddenly find that you can’t remember who you are. I had awoken one Sunday morning feeling unwell. I was due to go to work but felt sufficiently off colour to decide to stay home and after I had arranged with a friend to take my place at work I took it easy for the rest of the day playing some of my favourite music. In the evening I decided to write to my mother when, after writing a few pages, I suddenly felt very woozy and felt even more so when I couldn’t recall my name or recognise my surroundings or what I was doing. I must have had some of my wits about me as I decided if I was going to collapse, and I felt that I was, I was going to collapse in a public place. Fortunately almost immediately after entering the main street, I ran into a fellow student who greeted me by name and I regained my identity. I wasted no time in going around to the doctor, with whom I was registered. He was a very capable doctor, particularly in the surgical field, but had a reputation for taking drugs. He certainly showed fairly convincing evidence of that when in response to my knocking he opened the door clad in pyjamas. When he heard my story he pressed a couple of pills into my hand and bade me take them, go to bed and come back and see him in the morning, with the parting comment, “I’m a bloody sight worse than you are,” a statement I was prepared to believe judging by his appearance.

I changed my doctor next day when I had felt well enough to go to work. This one, after listening intently to my story and questioning me on my work and study programme came smartly to the point with the injunction, “No more study for the rest of the term." This didn't suit me at all. I had gone to him expecting him to prescribe some medicine that would set me up again and I was appalled at the thought of losing a whole term. We argued this point for some time and, after conceding that there were some subjects from which I could absent myself for a term, the doctor agreed that I could attend those with which I expected to work hard at provided I presented myself for his inspection at regular intervals. Then followed, during the remainder of that second term, one of the happiest experiences in what has been a very happy life.

During the previous two years on random occasions, which occurred not very often, I had been exchanging probing glances with a certain young woman whenever I passed her in the main street, the only place I would have seen her as my social life was virtually non-existent – one picture show in the first year and one dance in the second. I had no idea who she was but our glances would lock until we had passed. I considered her a stunning girl, good figure, good carriage, gorgeous looks and I guess that I fantasised about her quite a bit. A friend of mine, a fellow who used to eat at the same restaurant and because he had a girlfriend was more socially active, would often tell me about a girl who was very keen to meet me, a piece of information that I used to shrug off on the grounds that I was too busy, too fully occupied for girls to enter my life. A couple of days after the strange mental lapse when he observed that I had missed some lectures at the school and I explained the reason, his eyes lit up like neon signs and he said with some excitement, "There is a circus on here tomorrow night and I'm making up a party. Will you join it?" I like a good circus and after some hesitation, I allowed myself to be persuaded. When he was satisfied that I was not going to back off, he announced that he had a partner for me, the girl who had for some time expressed a desire to meet me. I half suspected, without daring to raise my hopes too high, that it would be the one girl for whom I had worked up quite a passion, but so it turned out.

We had a great time at the circus and, having found out that I did a bit of singing, she asked me if I would like to join the Choral Society of which she was a member and which was rehearsing Handel's Messiah for a subsequent ABC radio broadcast. I jumped at the chance and duly presented myself for rehearsal the following Tuesday night. It was a pleasant evening. The conductor, Eric Bray, an accountant from the local branch of the Commonwealth Bank, was very competent and the assembled choristers were fair average quality. After supper following the rehearsal, Leonie approached and offered to drive me home in

Daddy's car. "Oh, that's alright," I said, "I only live a short walk down the road but thanks for the thought," or words to that effect. I spent a fair proportion of the following week until the next rehearsal wondering how I could advance my cause with her. I experience acute embarrassment when I think of the repeat performance the following week when I again assured her that although it was only a short walk the exercise would do me good. The penny dropped the following day and I spent the next week wondering whether she could pluck up enough courage to again offer me her company in "Daddy's car". After the next rehearsal I think we were almost the last if not the last to leave and when in a very strained voice she made what now I am sure would have been her last offer, I had enough gumption to agree. We had many a good laugh about it later and for the next eight weeks or so we had a marvellous time. With the rest and the delightful company my health improved by leaps and bounds and when the exams came I was able to take them in my stride and pass everything very satisfactorily.

Then came the time when we had to call a halt. I had to devote all of my time to the third term leading up to the final exams. I explained all of this and Leonie took it very well. A few weeks later an American geologist, Stuart Millar, who had been a member of a team of geologists recruited in America by Dr Hugh McKinstry to reinterpret the structure of the Golden Mile and had been resident in Kalgoorlie for a year or so, called Leonie by telephone from New York to propose marriage. They had apparently been very close friends, sweethearts perhaps and, to my great sorrow at the time, she accepted. Events proved that she had made the best choice although I believed at the time that if I had made a similar proposal she may have accepted. When I called on her twenty-one years later in New York I realised that the lifestyle her husband provided for her in America suited her personality admirably and that had she been my partner I don't think she could have withstood the rigours of the road I travelled in building a career. Looking back, those eight or ten weeks were a gorgeous time. Perhaps it was the letdown that accompanied Leonie's departure for America and her marriage that prompted me to suspend my studies in 1939. Perhaps not. I was dissatisfied with the money I was earning and was offered an appointment to start up a newly built treatment plant at Spargo's. That I accepted the offer shows retrospectively a lack of judgement and a departure from my firm resolve to stay with my course and not be attracted away from it no matter what the perceived advantages.

I made some mistakes there, the main one being to reward the best of those who had performed well on the construction by training them to become plant operators to ensure continuity of employment for them, Had I been able to stay on the job twenty-four hours a day for the first three or four weeks I would probably have succeeded but some mechanical disasters occurred during the few sleeping hours I allowed myself. I could not excuse myself for these nor would the directors condone them and we agreed to part company.

Like most other treatment plants constructed at that time much of the equipment was secondhand. Where the machinery had in previous times seen numerous years of long service, some of the component parts were worn very thin and unable to take the strain of an overload. The most spectacular of the accidents was the result of a night shift operator allowing an overload of ground ore to build up in one of the thickeners which forced the thickener rakes upwards and wrapped them around the central drive shaft. It was a horrible sight. When a new mine starts up, invariably all available money has been spent and an immediate positive cash flow is imperative. An occurrence such as just described is calamitous. This was a decided setback to my career upon which I had just taken the first tentative step.

☆ ☆ ☆

The Spargo's experience was to some extent mitigated by a friendship that developed between me and Denis Martin, the mine engineer. I was twenty-nine and he was several years my junior, he may have been not much more than twenty, but he had moved around a lot for his age and had gathered a lot of very valuable experience. Besides his mechanical skills, he was innovative and a good lateral thinker. By the time of my departure he had reached a stage of overwork and disillusion, both generated by the very poor quality of the machinery, that he decided to leave with me.

We discussed possible future movements and mutually decided that a change of pasture would do us good and that we should seek employment either in New Guinea or Fiji. Neither of us had much money. I had just satisfied a substantial call on my bank account by my brother Frank. I don't know what Denis did with his, but he didn't have much and I decided that I should hold the combined purse and be treasurer. I applied a restraining hand immediately. Denis wanted to travel by first class but, after counting up the money and making a rough forecast of future needs, I bought third class tickets much to Denis' chagrin. On arrival in Sydney we booked into a hotel in Elizabeth Street about 2.00 p.m. and found that not only were our rooms not made up but were being used by prostitutes. This was too much for two boys from the bush to stomach and we crossed the bridge and skirmished around for a smaller, cheaper hotel. Luckily the first hotel we visited there, the Milsons Point Hotel, was leased by a woman from Boulder, WA. She was a motherly soul and offered us a large airy room with two single beds overlooking the harbour with full board for £2/15/0 per week each. The meals were first class home cooking and the atmosphere of the hotel wholesome and friendly. We stayed a week while we made up our minds on whether to go to New Guinea or Fiji. Fiji won. The *SS Niagara* was in and due to sail the following day. When we went to book our tickets, out of deference to Denis, I decided to relax our spending policy and travel second class. Bad luck; second class was fully booked, so according to our policy I booked two third class passages.

We decided to board early and stay on the boat overnight. Our cabin was right forward, abutting the crew's quarters which must have been right in the bow and with a chain locker on either side. We got very little sleep. It seemed to us that some of the crew spent the night practising raising and lowering the anchor while the remainder were practising their knife fighting skills in the fo'c'sle.

In the morning in the third class dining saloon we met the only other third class passenger, rather, I should say, "passenger travelling third class". She was a girl of about eighteen to twenty, very well spoken, quietly dressed and nice looking. I could see a competition looming up between Denis and me for her favours. We didn't have much time to talk to her other than to ascertain that she was returning to Vancouver, before the steward served breakfast. He was a villainous-looking chap who looked as though he hadn't fared too well in the brawling that had been going on all night in the crew's quarters on the other side of the bulkhead. His appearance from the feet up – dirty white plimsolls, dirty white trousers, a dirty white mess jacket, open, revealing a dirty singlet partly concealing a black hairy chest, surmounted by an unshaven face and a shock of black unkempt hair – was surpassed in unpleasantness by the meal he placed before us. Remember, this was breakfast. The meal consisted of a plate of thick, greasy curry of a colour so repulsive that it raised the thought that it must have been made with water from the "great grey greasy Limpopo". I tentatively stirred it over with my fork to get a look at the contents under the slimy covering of green sludge wondering what effect similar meals would have on my stomach when we were in the open sea pitching and rolling. I didn't consult Denis, I knew what his reaction would be, and excusing myself

I went straight to the purser to enquire if there were any unoccupied first class cabins. It was our lucky day, there was a two berth cabin available, the sole remaining cabin. I felt badly about leaving that very obviously nicely brought-up young lady to her fate but my chivalry did not extend to sharing those terrible conditions with her. We moved our belongings to our new quarters and it was like moving into a new world. There was just one penalty. Everyone dressed for dinner. These were days before I owned a dinner suit and Denis was in the same fix. However, we both had good quality dark lounge suits and we were placed at a table with two pleasant women who didn't mind. One of these was about my age and was an actuary in the Commonwealth Bank. The other was an almoner with a charitable organisation. We spent a lot of time in their company. I think they had both taken the trip in the expectation of finding a little romance. I teamed up with the younger one and quickly developed a nice friendly relationship with her. Unfortunately she didn't "turn me on" and it didn't get further than that. She must have been disappointed by my unresponsiveness. We corresponded for a short time after we left the boat at Fiji; she continued the voyage. In her first letter she commented that on the rest of the trip she found a male companion upon whom the tropical moonlight was not entirely wasted!

During that voyage I was given some very good advice by a master stevedore who had been holidaying in Australia and was returning to Fiji where he worked. On learning that we intended to work on the mines and hoped to save some money he told us that the team at the mine were very heavy drinkers. "If you don't join in you'll always be on the outer with them," he said. "Join in the first couple of times and then excuse yourself on the grounds that the doctor has put you off liquor. They will understand. There is quite a bit of venereal disease around and they will think you've caught it. After some time you will have been accepted and will be able to skip the rest of the parties." That is how it turned out.

During the course of the voyage we learned that before being admitted into Fiji, passengers travelling other than first class had to produce evidence that they were in possession of a minimum sum of £100 otherwise they would not be allowed legal entry and would be returned by next boat to Australia but that first class passengers were exempt. This was very fortunate for us as our money was nearly all spent and we were relying heavily on obtaining work quickly at either the Emperor or Loloma mines. We stayed in Suva over the weekend and on Monday took the bus around the island to Vatukoula. That trip convinced me that I was going to enjoy Fiji. The passengers on the bus were almost entirely Fijians and they sang island songs and hymns in beautiful harmony all the way. The passengers were constantly changing, the new passengers slipping easily into the harmony, softly, melodiously, no-one dominating the others. It was splendid.

We both got jobs on the Emperor Gold Mine the next day, Denis as engineering draughtsman and I as shift boss in what had been the pilot treatment plant. It was quite an experience. I commenced work that midnight after an all too brief orientation tour of the plant in the afternoon. The night shift crew was led by a boss boy called Taitusi who could not speak English and I could not speak Fijian. Within an hour of starting work Taitusi materialised out of the gloom with eyes flashing, arms gesticulating wildly and a stream of Fijian issuing from his mouth with all of the energy of a volcano erupting. It was fortunate that the English names of the equipment and machinery in use had been incorporated in the local language and I was able to pick up a word that sounded very like "thickener", so intimating by sign language that if he showed me I would follow. I located the offending unit, the pumps of which had stopped pumping, fixed it up and indicated that he should show me around the rest of the plant. The rest of the shift passed peacefully and the next day I bought a book on the Fiji language. Taitusi and I progressively taught each other the rudiments of our respective languages and slowly the rest of the boys picked up a bit of English too.

After we had found ourselves jobs on the mine and settled in we were invited to join a party to send off one of the staff who was returning to Australia to work at Wiluna. It was held in the open air, the group sitting around in a largish circle and the drink was beer. The privilege of buying a drink for the party was passing in rotation around in the group on the basis of a bottle of beer per head. The drinking was deep and fast. Before the first hour was up, I was half way through a second bottle and had two full ones in front of me. As the evening wore on I got further behind, encouraged all the time to "Slip into it, mate, and catch up," and assurances that after a few weeks I would be up amongst the best of them. I followed the old stevedore's advice.

I found the Fijians lovely people, warm, friendly and loyal. When I left Fiji to return to Australia to enlist, the entire shift went with me to Suva to see me off. They have probably acquired a veneer of sophistication since I worked with them. In those days they were simple and unspoiled. If they misbehaved at work and were fired as a result, it didn't mean a thing. They didn't hold it against anyone, they simply returned to their village and continued to cultivate their patch of garden which provided them with all the basic necessities of life. When they were working, after receiving their pay, they would ride the bus continually between Vatukoula and Tavua. Maurice Hedstrom had a store at each place, where they would progressively spend their money buying icecream, fizzy drinks, pocket knives, torches, ribbons and so on until it was all gone. Their pay was two shillings a shift for a plant hand and three shillings for a boss boy. I can't remember if pay days were fortnightly or monthly, probably the former, but whichever it was it was spent in one day.

The invariable conclusion to payday celebrations was a dance at night called by the Fijians a Tra-la-la. For many it resulted in the consumption of too much kava, the local drink, which taken to excess affects the nether limbs and the ability to walk. I was introduced to kava very early on in my job, probably after they had decided I was alright and should be invited into their social circle. For special tribal ceremonies the women would prepare the drink by arranging themselves in a circle around a large wooden bowl, usually supported on three wooden legs, called a tanoa, into which they would spit the fibre and juice of the roots of the yagona tree after they had thoroughly masticated it. The tanoa was then filled with water and the fibre which floated to the top skimmed off with a whisk of coconut fibre. It sounds rather unhygienic but alternatively it could be bought in packets as a dry fibrous powder which had been prepared mechanically and it was in this form that I participated in yagona ceremonies held during the shift at work. It didn't suit my taste at all. At work it was usually prepared in the enamel bowl in which samples of the cyanided pulp were gathered for testing. I always hoped that the dish was well washed out before the brew was prepared but judging by its taste and texture it might well have been a diluted portion of cyanided pulp I was drinking. Mixed by hand in the enamel bowl filled with water it was then bailed out in a half coconut shell and handed around the gathering. It had to be drunk in one draught whilst the rest clapped slowly in unison following which the vessel was refilled and handed to the next in line. I was told that many Europeans, long term residents of the islands, get hooked on it. It was definitely not for me. My participation was solely in the nature of a PR exercise.

Shortly before I resigned from the mine I was given, for experience, two weeks work in the gold room. It was interesting but smelting gold-zinc precipitate is fairly standard. A day before leaving, Guy Ditchburn, the mill superintendent, invited me into the strongroom. Piled up in the middle of the vault was a stack of four hundred ounce gold bars. These had accumulated from the practice of reporting only what was calculated from the weight of ore treated as recorded by the automatic weightometer at the assayed grade of the mill head sample less the gold lost in the residues. As the weightometer was overloaded most of the time it registered much less tonnage than was actually treated and the excess gold resulting was stored against a rainy day. I had been brought up on the basis that what gold was won over any given period was attributable to the people holding shares during that period. Management rationalised their action by pointing out the isolation of Fiji, and the possibility that in war-time, and the threat of war had been hanging heavily in the air for at least a year, the gold production from the mine could be interrupted for many months and in such an event the mine would be able to maintain its normal gold production for quite some time by drawing on the stockpile. I don't know if ever such an eventuality occurred but it must have been a great comfort to the management to know that hoard existed.

I was still working in Fiji on the Emperor Gold Mine when Hitler invaded Poland. Having read during the previous few years of his walkover conquests of other European countries the news of his Polish invasion was not unexpected and we had discussed what we would do if war was declared. When Australia, following Britain's example, declared war on Germany the following day, seven of us, members of the mine staff, all single men and hence footloose, decided to quit and head for the mainland to enlist.

The word had got around that in the event of war the mine management had lined up a deal with the Fiji government pledging a company of volunteer soldiers to form a homeguard from the ranks of its employees in return for assured oil supplies for mine power generation. None of us relished the thought of spending the war years in Fiji and as the *SS Niagara* was calling in at Suva on a scheduled return from the USA to Australia, we booked our return to Sydney on it. This was quite a blow to management as there weren't all that many spare staff on the payroll but as the boat was leaving within a couple of days we got safely away before the manager could organise any restraining action.

Because the pilot plant worked seven days a week I didn't have much spare time in which to dissipate my salary and when I left I had built up a reasonable bank account. As may be imagined, I travelled first class on the return trip.

The army was very choosy in those days and all of us were found unacceptable for one reason or another. Although I had 6/6 vision in both eyes when wearing glasses the recruiting sergeant in Sydney insisted that I take them off for the test and promptly failed me. The others suffered a similar fate for different reasons. One of the party, a qualified fitter and boilermaker welder, very keen to join the Armoured Division, was rejected because he was one half inch short of the required five feet six inches! Blocked from doing my bit for King and country I returned to Kalgoorlie with the intention of getting a job there and continuing my part-time studies at the School of Mines.

Chapter 6

When in 1939 I arrived back at Kalgoorlie from the Fiji jaunt I took a bed-sitter in the house of Richard (Dickie) Moore, the mayor of Kalgoorlie. He was a wonderful old fellow, a blacksmith by trade in which he had his own business. He was so successful as mayor, mainly through his commonsense, practical approach and decisiveness, that he held the position for twenty-nine consecutive years. He was a great champion of Kalgoorlie which had a widespread reputation for being dry, dusty and insufferably hot. It was certainly very dusty in those days as the depredation by the woodcutters who supplied the gold mines with mining timber and firewood for boilers and ore roasters had denuded the mature timber and even immature regrowth for a radius of a hundred miles.

Dickie, who was years ahead of his time, after the war established a green belt around Kalgoorlie with plantings of local eucalypts and also planted thousands of street trees in the town. The basaltic soil of the goldfields, given supplementary watering, is a fertile base for any type of vegetation particularly for jacarandas which are very popular with householders and are extensively planted in domestic gardens.

When, during a mayoral visit to Bridgetown, the local mayor took Dickie to a lookout, extolled the beauty of the landscape of lush pastures and had a sly dig at him with the question, "How would you like to live here Mr Moore?" Dickie's reply was swift in coming, "I would love it if I were a cow." Dickie was rewarded for his years of service to the community with the award of a knighthood.

During the short interval before picking up the job at Coolgardie I met Dorothy Jolly, the girl who later became my wife. Her parents were long time residents of the goldfields, her maternal grandparents with their family having lived in Coolgardie from before the turn of the century. Her grandfather was a winder driver on the Red Hill gold mine at Kambalda which was later to become the centre of the nickel industry in WA. He would ride his bike out on Sunday afternoons, forty miles along a bush track to the mine and return to Coolgardie on Friday after work to be with his family for the weekend. They certainly bred men tough those days!

Dorothy grew up in Kalgoorlie, attended high school there and then went on to the University of Western Australia in Perth where she graduated in Arts majoring in Psychology. She joined the Education Department and as her first posting was to Coolgardie at the same time as my appointment as mill superintendent to the local gold mine Phoenix Gold Mines NL it was in the normal course of events that we spent a lot of time in each other's company.

When I took up the position of mill superintendent I purchased a small Vauxhall car called a Wyvern Caleche. It was a sporty little thing with a fabric top that folded back and a windscreen hinged at the bottom so that it could lie down horizontally and I am sure it added to whatever other attractions Dorothy saw in me. It wasn't long before it became a regular thing to take her home on Friday evening and return her on Sunday, an arrangement that pleased her parents so much that I was welcome to stay with them over the weekend.

Under these conditions friendship quickly blossomed into something more emotional and the day came when having ascertained from Dorothy that marriage wasn't a bad idea I fronted up to old Jolly and asked his blessing to this proposal.

This formality having been successfully negotiated, the next step was to announce our engagement in the press. Regrettably I didn't have any loose change at that time to buy an engagement ring and I quickly came to learn the importance a girl attaches to confirming the media announcement by displaying an engagement ring and that the larger the diamond the more effective is the ritual considered.

This put me in a very embarrassing position. The down payment on the car had used up the money I had saved while employed at Fiji and, after paying for my board and the instalments on the car plus running expenses, there wasn't much left out of my monthly salary of £40. I stalled for a long time, too long for Dorothy who, being a girl of action, decided to take the matter into her own hands and arranged for a Perth jeweller to send some samples up by mail from which she could make a selection.

The arrival by mail of a packet insured for £105 containing three rings each priced at £35 took me unaware but realising that she meant business and that I could no longer equivocate I had to come clean and tell Dorothy that I didn't have the wherewithal and that some time would elapse before I did. She took it well and endured for quite some time the indignity of not having an engagement ring to compare with those flaunted by her friends.

It wasn't long before she needed tangible proof that she was a girl of good moral standing. Social activities in Coolgardie were conspicuous by their almost total absence. I think there was a movie show once a week and when my duties at the mine prevented me from making my normal visit, there was very little for her to do other than read a book. One such evening she was reduced to talking to an ageing Lothario of about sixty-five or so, a mining engineer visiting Coolgardie and staying at the same hotel.

Dorothy

Coolgardie boasted two hotels in those days, the Denver City and the Railway. It was winter, a season when it can get quite cold on the goldfields, Dorothy commented in the course of conversation that she suffered from cold feet and had been unable to buy a hot water bottle in Coolgardie and yes she would be pleased to borrow a pair of the old gent's nice thick woolly socks until she could buy one in Kalgoorlie. During the night they proved to be too hot. She kicked them off and forgot them in the morning. The housemaid found them when she made the bed and when Dorothy came home found them neatly folded on her pillow displaying the old gent's name on those very clearly marked personal name tags marketed by "Cash"! Attitudes to extra-marital sex were vastly different then to now and the gossips made the most of what they inferred had occurred.

The mill superintendent's salary was higher than I would have expected to receive for a position in Kalgoorlie and I deluded myself into thinking that I could commute between the two locations when the school resumed in February. It did not work out that way.

The position I accepted with the Phoenix Gold Mine NL was a company within the Bernales Group. Claude Albo de Bernales, to give him his full name, a dominating figure in Western Australian gold mining throughout the 1930s, was a man of compelling appearance, tall, handsome, always immaculately groomed, full of charm and, because of his relaxed ethical standards, which were very similar to some of Australia's successful entrepreneurs of the 1980s, he became very wealthy over a very short period at other people's cost. He had become a local legend, although because a history of his life has not been published the legend is beginning to dissipate. It may be that a comprehensive biography of his life has never been published because of some discreditable business ventures that he either initiated or became associated with.

Claude de Bernales arrived in Western Australia in the early days of the development of the eastern goldfields of WA. Never one to soil his hands, he first featured as an agent for mining machinery. The distances between towns and between mines were substantial, the main roads were unmade dirt tracks usually with a cycle pad alongside. Claude used to travel these by bicycle, making the run from Kalgoorlie north to Leonora and Lawler's twice a year, visiting the mines that were off the beaten track, a practice that earned him many customers and loyal supporters. Riding a bicycle on a dusty track on a hot day is sweaty work but Claude would always make himself presentable, donning a clean shirt after a quick wipe over with water from his waterbag. When the mining industry began to wane during and after the Great War of 1914 - 1918 he acquired most of the mining equipment of the many mines that had folded from one cause or another, either from ore cutting or the high inflation that followed the war, and held it under machinery title. Later he acquired, by marriage, control of the Kalgoorlie Foundry.

In 1932 America devalued its dollar in terms of gold from US $22 per ounce to US $35 per ounce which initiated a second gold boom, sparking great activity in Western Australia where mines that were unprofitable at the old price offered prospects of again becoming profitable.

The eastern goldfields of WA, particularly Kalgoorlie and its environs, became a hive of activity. Mining equipment was in great demand, particularly second-hand equipment which was immediately available and over which Claude had a monopoly. His modus operandi was to sell this previously obsolete equipment to the Kalgoorlie Foundry where it was reconditioned, given a heavy coat of paint and on-sold. Very frequently the customer was a company which Claude himself had floated on mining leases he had taken up.

Phoenix Gold Mines NL was just such an outfit at Coolgardie, which was based on the reefs that Ford and Bayley had found in 1892, a momentous discovery, which ushered in a period of discovery lasting several years and included the Kalgoorlie field found in the following year. "Outfit" is the operative word as the mine had been outfitted with plant and equipment which had passed through this sequence of events. The treatment plant comprised a twenty head stamp battery, a tube mill, a pair of reciprocating rake classifiers, a pre-agitation thickener, six agitators and four decantation thickeners, all of which were in pretty fair condition.

The power plant however was not. Originally installed at the Sand Queen at Ora Banda it had been reconditioned at the Kalgoorlie Foundry. It comprised a number of large, slow-revving internal combustion engines with a combined rating of seven hundred and fifty hp powered by gas produced by the partial combustion of wood in gas producers. The painter at the foundry had done an exceptionally good job on them and they made a handsome show installed in the power house and ran sweetly on "no load" tests. However the eventful day came when the mine was able to sustain continuous production and with the ore bin at the plant full of ore and shareholders doubtless full of expectations of a return on their investment, the power plant had to perform under load. It was indeed an eventful day, marked by twenty-seven power failures in the twenty-four hours. Every time the power went off the battery hand had to make a frantic rush to chair the stamps before the cam shaft stopped turning. At the end of the first week of this performance he had become very dexterous at this particular task, urged on no doubt by the knowledge that those that he couldn't pick up while the cam shaft was still turning would have to be laboriously raised by block and tackle.

When the tube mill ceased to revolve, the load of finely crushed ore it had received from the battery and was in process of grinding still finer, disgorged onto the mill floor to confront the plant hand when he returned from outside the mill building where he had been frantically raising the rakes in the agitators and thickeners to prevent the suspended slurry from settling around them. It wasn't long before the accumulated slurry on the mill room floor was knee deep. When it began to overflow into the operator's gum boots, activities had to cease until it was hosed into the mill floor sump and pumped into the classifiers. Usually the mill hands would have just got all of the stamps on the chairs and raised the agitator and thickener rakes when the engineer would get power back into the system and the circus would start over again.

After a couple of days of this merry jig, which happily saw the incidence of these stoppages reduce slightly in frequency, it was decided by the foundry engineers that a close examination of the pistons and cylinder liners was justified. This indicated that the mechanite alloy from which they had been cast was "green" and had warped. They were therefore taken back to the foundry, the cylinder liners honed and the pistons turned and restored to circular form from the oval shape they had assumed, a process that had to be repeated several times as it only reduced the frequency of power failure and did not eliminate it.

During this very trying period, I think we had been going about a week or ten days, the managing director appeared on the scene to tell me that he was bringing the full Board to inspect the new installations two days later. As the date and time of this visit had been confirmed with the directors he flatly refused to put it off until the power house was running efficiently, apparently he was unaware of the scale of the problem until I told him. He then agreed to keep a watchful eye on me during the course of the inspection and at my first sign of looming trouble to hustle the party out of the treatment plant into the waiting cars and away.

There was insufficient time to take the cylinder liners and pistons out of the engines, up to the foundry for attention and re-install them before the impending visit, and other means had to be adopted to ensure that the inspection was accomplished without a power failure and the attendant chaos which would result in the treatment plant. On the day of the visit the engineer, Alf Horton, and I decided that I would shut the treatment plant down at 8.00 a.m. to allow time for the engines to cool down and for a good plant clean up. The engineer would start his engines, all seven hundred and fifty hp of them at 10.45 a.m. The power house and general office were about five hundred metres from the treatment plant. As soon as the visiting party had inspected the power house and departed for the mill, a cleaner would hoist a white rag on a stick and wave it vigorously to warn the mill hands who would then activate the plant.

By the time the cars were parked and the visitors entered the battery house everything was operating smoothly. I took up a strategic position from which I could see the electrical distribution board, in particular the voltage meter, and command a general view of the interior of the mill room. The battery hand gave a faultless demonstration to the admiring directors of putting the battery stamps on the chairs and taking them off again to resume crushing, a not surprising feat considering that he had had an awful lot of practice in the preceding two weeks. The look of satisfaction and pride on the faces of the directors as they admired the array of smartly painted, well cleaned, reconditioned used plant was enough to warm any shareholder's heart.

The engineer must have decided to give himself a little leeway as he was churning out the power ten volts above the normal four hundred and forty volts. By the time the party returned to the mill room, after admiring the agitators and thickeners and a couple of dead cows that had drunk deeply of the solution in the new residue dam and had paid the ultimate penalty of quenching their thirst with cyanide solution, the voltmeter was registering four hundred and twenty volts and falling perceptibly. A hurried frantic gesture to the managing director and he smartly ushered his party outside. The cars had just moved out of earshot when the needle on the voltmeter went into rapid decline and the whole outfit stopped. The regular routine of power failures continued and the Board sat down to lunch happy in their ignorance of what a dreadful mess the power plant was. Of course it could not continue in that mode. A one hundred and eighty hp diesel was brought down from another of Claude's mines and hastily installed with the aid of quick setting cement. When it was started a few days later it handled the total load without a hiccup and ran uninterrupted for the next twenty-eight days when it was stopped for routine maintenance. As far as I know it continued to operate satisfactorily until the mine shut down during the 1950s. In the meantime the seven hundred and fifty hp of assorted gas engines rusted quietly away.

That episode, which lasted about three or four weeks, speaks volumes for the quality of the working man of fifty years ago. It was a lousy job but to my recollection they all stuck at it until the faults were ironed out and operations settled into a regular routine.

Well, hardly an unbroken routine. Bill Thirloway, the mine manager, was well versed in the ways of the average chap working in the gold mining industry and, knowing the propensity they had for pocketing any gold they could get their hands on unobserved, had every place where gold concentrated either by design or by chance protected by steel mesh and secured under lock to which I held the key. This caused no problem during day shift but the plant operated twenty-four hours a day, seven days a week, and every time something went wrong at one of these locations I had to be tracked down to unlock the mesh protecting the particular area, supervise the adjustment, repair or replacement and re-secure it. This used to occur most frequently during sleeping hours prompting me to protest most vigorously to the manager that this procedure was entirely unnecessary as the workmen were totally trustworthy. It was to no avail. The manager, a wise old bird who had come up through the ranks, and who mistakenly thought my name was Brogie, knew better. "Where gold is concerned, Brogie, never trust nobody," a position from which I was unable to budge him except where I was personally concerned and that was only after a probationary period.

In the Phoenix plant, after the ore was crushed by the battery and the coarse gold released in that process recovered by amalgamation with mercury on copper plates, it was further ground in a rotating cylindrical mill to a size at which the fine gold remaining could be extracted by agitating the ground ore in a solution of cyanide. The gold-bearing solution was then separated from the slurry, clarified and the gold recovered by precipitation on zinc shavings placed in specially designed multi-compartmented tanks. The gold-bearing solution flowed from one end of the tank to the other progressively depositing its gold content en route. Most of the gold precipitated in the first two or three compartments causing the zinc to disintegrate and the resultant zinc-gold slime to sink to the bottom of the box. By the time the solution had passed through the length of the box, all of the gold had precipitated. Progressively the zinc was moved forward along the boxes and new zinc added at the end at which the now barren cyanide solution was discharged back into the circuit. The rich gold-zinc slime was then removed and treated with strong sulphuric acid to dissolve the zinc. The remaining gold-enriched sludge was washed with water, dried and roasted in air to oxide any remaining zinc, following which the dried roasted sludge was smelted with fluxes which slagged off most of the remaining base metal impurities and a bar of gold was produced.

I was disconcerted when the first clean up of the zinc boxes was started to find the manager on the job and scrutinising every move. I became more and more annoyed as the day wore on to find that he stayed there all day. Finally, when the treated but not yet smelted sludge was weighed and locked in the plant strong room, he insisted on keeping the key. The following day he was on the job again and didn't vacate his post until he had the weighed bar of gold firmly in his hand and was on his way to his office to lock it in his safe. I was incensed at this obvious lack of trust in me and told him so with some force. It was then that he came out for the first time with his famous dictum concerning trusting nobody.

For the next three or four clean-ups and recovery of the gold I became accustomed to his constant surveillance and it ceased to bother me. Then on one occasion when the gold had been poured into the ingot mould and was tipped out, on cooling it was found to have a layer of white metal about half an inch thick on the surface. "What's that, Brogie?" A white metal bearing had melted a few days earlier and being fairly certain that it had somehow reported in the gravity circuit and been amalgamated with the gold, I gave this explanation and told him that if the bar was put back into the still red hot furnace, the white metal would melt at a much lower temperature than the gold and would run off and collect in the bottom of the crucible. Such proved to be the case and after the bar was cleaned up, weighed and stamped, the manager went away quite content.

As there was a very good chance that some gold had become alloyed with the white metal and as the gold bar weighed a bit less than I had calculated, I decided to pour it off into some disused assay moulds for further treatment. There were then ten small conical ingots of what at first glance were white metal. Part of my normal duties included assaying the ore samples from underground, a task which was seldom completed before the end of the day.

The cupel furnace was still hot enough to treat one small ingot at a time and having improvised a large cupel using a small tin as a mould, I encased the ingot in sheet lead, put it into the cupel and put it in the furnace. I was delighted to find when all the lead had been absorbed or volatilised, a beautiful spherical button of gold remained which on cooling weighed one and a quarter ounces.

I took this over to the manager and nonchalantly threw it on his desk telling him in answer to his suspicious query, "Where did you get this Brogie?" that I had cleaned up some slag and other potentially gold bearing rubbish that was lying around. "Is there any more of it, Brogie?" "Oh yes, there could be a bit more." This ceremony and dialogue was repeated almost identically for another nine days until the last of the ingots had been similarly treated. By this time the manager was almost off his head with frustrated rage. Upon handing over the final button of gold I told him where they had come from. He gave me a hard look, obviously deciding whether to give me a blast or not and then turned away with a non committal grunt. From that time on he took no further interest in the regular recovery of gold from clean-up of the zinc boxes.

But the old chap was right. A couple of years later, when I was in the army, he sent me a paper cutting from the *Kalgoorlie Miner* and a pithy little message, "Where gold is concerned, never trust nobody." One of the battery hands, a person whom I had always held up as an example of perfect trustworthiness, had been taken off the Trans Australian Railway by the gold detection staff with seventy-five ounces of gold in his possession. I made a point of calling on the manager when I returned to Kalgoorlie at war's end to admit that he was right. "Ah Brogie," he said, "With gold, never trust nobody." Advice I have followed ever since.

In due course the treatment plant settled into a monotonous routine and at the end of the period of "Phoney War"in 1940, I decided to try to enlist again but was thwarted in that effort also.

Australia had a pressing need for hard currency to finance its war effort and, though gold production was small in relation to today's production, it was considered important by the Federal Government and employees of the industry were prohibited by law to relinquish their employment to join the forces.

It was probably just as well for me that I was again excluded because those who were accepted at that time were eventually posted to Singapore and became prisoners of war when the Japanese over-ran that supposedly impregnable bastion. A friend of mine who did enlist at that time participated in the Japanese celebrations held to honour the occasion. He had his head cut off with a sharp sword.

However, excitement of a different sort was waiting around the corner.

☆ ☆ ☆

The Bernales Group, which operated under the name of Australian Mines Management and Secretariat (AMMAS), had among its portfolio of operations a small company with the imposing name of the Beryl Gold Mining Corporation whose major asset was a group of leases at Kundip in the Shire of Ravensthorpe.

Early in the century Ravensthorpe had been a small thriving copper producing centre where a German company built a copper smelter and toll smelted ore for a cluster of small mines which developed around it. I think the main mine was financed by German money and Germany was the destination of the copper produced.

It operated until the first world war and then closed down. The Australian market was well supplied from other domestic sources and the international market was not available to it because of the shortage of shipping and, I imagine, the likelihood that a consignment of copper might finish up in Germany.

Claude de Bernales had acquired a number of leases (mineral, machinery, tailings) at a place called Kundip, situated about ten miles south of Ravensthorpe, upon the prospects of which he floated the Beryl Corporation. From the residue of the proceeds of the float after he had taken his vendor's reward, some exploratory drilling had been done. One hole made a wide high grade intersection of copper-gold at a depth of about four hundred feet which with the results of some of less successful holes was considered to be sufficiently encouraging to sink a shaft to investigate what was there. When the high grade intersection was exposed by underground development it was found to be a discrete blob of ore about the size of a small room with no extensions either vertically, horizontally or any other plane. A blob indeed, especially for the shareholders. Some further exploratory work was done and a small underground operation was established. When I became involved, the staff for whom some houses had been built had shrunk to about six people, two of whom were gouging out ore on tribute.

Over the years a not inconsiderable tonnage of mill residues had accumulated, spread around the district over a radius of about three kilometres, much of which, by combining the values of the contained gold and copper, was of economic grade.

It was the intention of the AMMAS Board to process some of the higher grade material, which averaged three and a half per cent of acid soluble copper and twelve and a half grams of gold to the tonne, and a sample had been tested by a contract metallurgical laboratory in Kalgoolie to determine the conditions under which it could be profitably treated. This was at a time when laboratory services were much in demand. Later events led me to suspect that either the sample was not representative of the total material or the testing and the deductions drawn from the results were not conducted to a satisfactory standard.

I must have made a favourable impression at the Phoenix mine, for the managing director offered me the job of constructing a small treatment plant and power house and the management of the operation when it was completed.

For a young budding engineer this was a wonderful opportunity to demonstrate his ability and I accepted on the spot. Within eighteen months it was to provide an experience in improvisation that in other circumstances would take decades to acquire.

The time was about September 1940 and petrol rationing was in force. For convenience on site I wished to take my small Vauxhall convertible. The distance from Kalgoorlie by the shortest route via Esperance was about three hundred miles. A friend managed to get me a twelve and a half gallon drum of petrol and, with full tank and some ration tickets in reserve, I set off.

The trip proved uneventful. After leaving behind the eight feet wide strip of bitumen sealed road from Kalgoorlie to Coolgardie, the road was gravel sheeted to Esperance. From there to Ravensthorpe, which was rarely travelled, the road consisted of two well rutted wheel tracks in sand plain. In some places the ruts were so deep that with a low slung vehicle, and my car fitted that description, there was a real risk of bellying the car sump on the central mound. There was no near roadside habitation in those days nor any roadside supplies of water, and to safeguard against the risk of being stranded by breakdown or any other cause, it was customary to report to the police station on departure and give the police officer the expected time of arrival, information which he would telephone through to his counterpart at the other end who, in the event of a long delayed arrival, would proceed to look for the missing vehicle.

The infrequency with which that one hundred and ten mile road was travelled lends some humour to an event that occurred along it three months later. Dorothy had arranged to visit Kundip over the Christmas holidays, an event for which I had prepared by engaging a live-in housekeeper for the duration of her stay, a precaution which did not stop my team of dedicated workers from referring to her as, "The boss's wife on appro"! It seemed to be a long age since I had seen her and I was waiting for the time when I could hold her in loving embrace. This was a time long before it became common practice to give a display of passionate love-making in public as is now quite common. I met her off the Kalgoorlie train at about 9.30 a.m. and after eating breakfast at the hotel and spending a couple of hours looking around the sights of Esperance, took off for home. I had the fabric top of the car folded down and the windscreen in the horizontal position and we glided along wrapped in young love's dream. However, even in that condition, hunger still has to be appeased and on finally encountering a shady tree on that otherwise practically tree-less road, I pulled up and as a preliminary, an aperitif you might say, I turned around, engulfed Dorothy in my arms and was engaged in giving her a long passionate kiss when a bewhiskered old gent in a clapped out utility appeared out of nowhere. Why I didn't hear his vehicle I'll never know but I quickly became aware of his voice, loud and raucous, "What the bloody hell are you silly buggers doing there. Get off the bloody road and let me pass." Before he had backed off onto the track he had emerged from to let us pass, there was more of the same. I think he must have been envious! On subsequent trips I didn't meet another vehicle.

However, that initial trip was done on time and on arrival at the mine on first acquaintance I found the core of people left to man the lease after the failure of the underground mining operations to be a nice bunch and, later on when the action started, they proved to be real gems equal to any occasion, never overcome by adversity and intensely loyal.

Although most of them were rich in years and, apart from a blacksmith, had no specialist skills, they could turn their hands to most things and were always prepared to "give it a go". I recruited a carpenter from Newdegate and later on when the diesel engine and alternator were to be installed, Frank Robertson, the son of the AMMAS accountant in Perth who had skills in this area, joined the team.

I find that people today are unaware of the tremendous technological advance in all engineering areas that was initiated by the second world war and has accelerated ever since. The construction of the Beryl plant was accomplished without mechanical equipment of any significance. Manpower was the main energy source. Excavations were dug by pick and shovel and the spoil carted away by horse-drawn dray. Concrete was mixed by hand and moved to the formwork by wheelbarrow. Today ready-mixed concrete can be delivered to the site and where necessary pumped to inaccessible places, and mobile cranes of varying capacity are available for hire by the hour or day to raise materials and equipment into position. In those days small loads were manhandled into position and for larger loads a stick was rigged. This was usually of oregon and up to thirty feet long. After standing it and guying it for stability the loads were raised by block and tackle, chain block or winch and manoeuvred into position by slackening the guy ropes and leaning the stick to position its suspended load over where it was to be placed. Sometimes it would be necessary to rig two sticks and transfer the load in mid air from one to the other in order to place its load at its final location. This could be a nerve-wracking experience. Because we had no certified rigger on the job, as much as possible of the lifting and positioning was done manually.

I recall most vividly erecting the steel columns supporting the power house roof. Having been dumped by the carrier some distance from the power house site they were carried manually to their respective concrete footings by three massively built Italians and one rather weedy Australian who proved to be deceptively strong. After aligning the base with the bolts in the concrete footing and tying a long rope to the top of the column, each one was raised into position by three of the party lifting the top end and "walking it up" to a height at which the fourth man could get some purchase on the rope. As it was raised higher all the men pulled on the rope until it reached the vertical. There was always great haste in dropping the rope and grabbing the column to stop it coming over too far. Until we got the knack of it the columns teetered perilously but luckily none fell. We were proud of our performance, standing all of the columns, I think there were twelve, in half a day. This procedure was typical of the construction phase. Manpower and ingenuity replaced today's wide range of specialised equipment available on hire for any job.

The construction progressed on time and as far as I know at an acceptable cost. I didn't have a construction budget to work to and the accounts were paid from Perth.

One of the final tasks to complete the construction programme was to install a pump on a mine shaft situated about a mile away from the treatment plant, and to lay a pipe over the intervening valley. During an on-site visit by the managing director and the metallurgical consultant, it was revealed that the money for the project was all spent and none was left to pay for the pipe. The consultant, a resourceful fellow wherever he sensed the possibility of making some extra money, dredged up from the dim recesses of his memory the whereabouts of some six inch diameter wood stave pipe in the Kalgoorlie area which, if still there, could be obtained at a very low price. I think a shilling a foot was mentioned but that seems hardly enough to include the consultant's commission for bringing such a bargain to general attention. The MD was ecstatic and decided on the instant that such a bargain should not be left lying around for someone else to snap up and made an on-site decision, sight unseen, authorising its purchase and transportation to the site forthwith.

In the newly discovered knowledge that all of the company's money was spent and that our own salaries and wages could not be paid until we generated money from operations, I and my band of workers were all agog for the pipe to arrive.

When about one week later a semi-trailer arrived on the job heavily laden with what appeared to be a load of junk, we jumped to the conclusion that the driver had lost his way and had come to us seeking direction. He was however quite firm that the Beryl Gold Mining Corporation was his destination and that his load was for us. On this information we examined the load in detail. It consisted of a tangle of large diameter long steel spirals of one quarter inch mild steel and long laths of wood, some broken, others whole, all exhibiting that beautiful silvery patina developed by timber that has been in the sun for decades and that characterises your genuine antique. We were on the verge of directing him to the rubbish dump when one of the Italians said, "By Christ, boss, it's the wooden pipe you've been waiting for." This chap had previously displayed a rich sense of humour and we joined in heartily at this joke. Then another chap, more perceptive than the rest of us, said with an air of conviction, "You know, boss, he's right. It's your wood stave pipe alright. " He was of course correct. In fairness to the consultant I must assume that they were in pipe form when they left and that they had all collapsed due to the shaking they had received in traversing about seven hundred kilometres of heavily corrugated gravel road.

Then followed long days of reassembly. The pipes had male-female ends and a lot of the male ends were broken. Fortunately there was more pipe than was required and by careful selection we managed to reconstruct enough for the job. But what a task it proved to be. Supported on three trestles rigged up for the purpose, the wooden segments were carefully arranged in circular form inside the wire and a uniform length for each maintained to ensure the ends were flush. Anchoring it at one end with a staple, the wire was pulled around with claw hammer and pliers and secured at intervals with staples in the wood that had been dessicating in the weather for goodness knows how many years until it was bone dry and ready to split as each staple was hammered home. The trick was in doing it so that the preformed spiral of quarter inch round steel wire was neither too long nor too short when the end of the pipe was reached. Laying the pipe when the job was completed was relatively easy. It was when we started to run the water through it that the fun and games really started.

The pump, installed about one hundred and fifty feet below the shaft collar, was a six inch diameter three throw reciprocating plunger type. It delivered a good steady stream of water which on start-up all leaked out from between the staves of the first pipe. As this pipe was a fair sample of the condition of the rest, it was a reliable indication of how they would all react. Then began the search of the countryside for any sort of material that would seal the leaks until the wood swelled and became water tight. Sawdust was the most desirable sealant but not much of that was available and was successively replaced with bran, chaff and horse manure. By building mini dams around each pipe to soak the pipes externally also, the water was slowly coaxed down the valley and up the other side. It was two weeks before the water started to flow into the head tank and we were able to commence operations.

At Easter 1941 Dorothy and I were married. I drove up from the mine to Perth on Good Friday, the ceremony was held on Saturday and we returned to Beryl on Easter Monday, arriving home at 3.00 a.m. Apparently it was the custom in those days to build the boss's house right where the action took place and there were all too many nights when a plant hand would knock on my bedroom window to get me out of bed to fix something or other.

The operating phase proved to be as challenging as the start up of the Phoenix plant but it lasted longer. In fact modifications to the plant continued throughout the time I was there, in a vain endeavour to achieve the throughput and recovery for which the plant had been underdesigned.

The most interesting of these was the occasion when we converted a thirty feet diameter head water tank into an agitator.

There was no money in the tin to buy a standard superstructure or indeed the materials from which to build one on site. Throughput was far below the designed capacity and hence gold recovered was less than Head Office expected and I was constantly receiving telegrams from them asking for more than I had to give. In the early stages these were peremptory demands to "Send more bullion!" but after a while became quite supplicatory – "Please, send more bullion". I am very sensitive to other people's distress and I found these plaintive appeals quite worrying. When I was smelting to meet these requests and knew that the outturn would fall short of what was required I often wondered whether I should toss my gold pocket watch into the crucible, an urge which fortunately I resisted in favour of a longer term solution.

The old German copper smelter which still etched its outline starkly against the sky had in its structure some very large (fifteen inch by fifteen inch by thirty-two feet long) jarrah members. Jarrah is a very hard, heavy and strong construction timber and if I could get one of these down and transported to the plant site where there was an old circular saw and saw these lengthwise into two beams I would have the main members to support the agitator mechanism.

I enlisted the help of Wally Onley, the contractor who hand-loaded and carted fifty tons per day of tailings into the plant for treatment. He was just the man for the job, deep chested, broad shouldered and very strong. The beam weighed a bit over a ton and could only be got out by manhandling. We took his truck and an assortment of tools, large spanners to loosen the holding down bolts, cold chisels and a flogging hammer in the event that the nuts were rusted on, a couple of crowbars and some rope. We got one of these sticks down from its elevated perch and loaded without inflicting any great damage to Wally's truck and managed to get it reasonably balanced, although there was a strong tendency for the front wheels to lift off the ground which made steering interesting especially when going down hill, and safely to the saw bench. The undercarriage of two mine trucks provided the chassis for two trolleys and when a short length of light rail was laid fore and aft of the saw we were ready to go. The blade of the saw must have been about sixteen inch diameter as it was just possible to cut the stick of timber, which had shrunk to fourteen inches, in two passes.

The saw was probably acquired to perform lesser duties than it was now asked to perform and the five hp Lister kerosene engine that powered it was like a willing but tired old man worn out by years of work and neglect. It put up a good pretence of being capable of doing any sort of job as it putt-putt-putted away under no load but when the timber was gently eased on, the saw issued diminishing screams of protest for two or three seconds until the inertia stored in the flywheel was used up. The putt-putting of the engine almost ceased before again slowly gathering momentum, challenging me to give it another go. It took about eight hours to make the first cut, and after turning it over we kept going until we finished it. It was just another twenty-four hour shift which had become a regular feature of the job.

The two beams which resulted were just long enough to span the tank and sit firmly on the trestles that the carpenter made. There was still another problem to be overcome. The agitator shafts in the acid copper circuit, where the new agitator was to be put into service, were four inch diameter stainless steel, very expensive, and for which we had no money even if under war-time conditions it was available. A less costly expedient was to have a crown wheel cast at the Kalgoorlie Foundry with an eight inch square opening to take an eight inch square timber shaft. We used karri timber for the shaft which is fairly tough and long grained. By careful nursing it did the job!

Kundip is about sixteen miles from Hopetoun, a little settlement on the south coast that was once a stopping point for the coastal shipping service, but ships seldom, if ever, came to Hopetoun when I was there. It had become a small retirement village for people from the surrounding farming community and a few astute outsiders who knew a peaceful haven of tranquillity when they saw it. People lived for ever there. The postmaster and his wife, both over ninety, regularly attended the dances that were occasionally held. A local legend had it that they would join in that stirring dance, the Lancers, but I can't vouch for it. Mrs Postmistress was a keen fisherwoman who would go down to the jetty almost every day, winter or summer, rain or shine, and throw out a line. To the intense chagrin of the local men who had been sitting there for hours and catching nothing, in a surprisingly short time she would haul in a half-dozen herring or skipjack, whatever was the fish of the day, and smiling teasingly at the men, retire, leaving them if not frothing at the mouth, certainly gnashing their teeth. Sad to relate, in the year or so that I was there I managed to get to Hopetoun only twice for day visits.

It was a relief when in November 1941 the Japanese having bombed Pearl Harbor, the USA entered the war and gold mining was no longer classified as a protected industry and I was free to enlist. It was poetic justice that the chap who had done the laboratory tests on which the plant was designed was sent down as my replacement. It interested me to learn that he gave up trying to treat the rest of the tailings but bought in from around the district other tailings with less difficult characteristics.

For my part, I think taken all round I enjoyed the experience. They were a great team of chaps who threw themselves into the problem of trying to get the system to work with as much zest as I. I regretted being associated with a failure but in the excitement and disruption of the war it passed unnoticed. When I completed the Metallurgy Associate Diploma course at the School of Mines I submitted my thesis on the Design and Operation of the Beryl Retreatment Plant. Years later I went back with Dorothy to Kundip on a sentimental journey. All the houses had been taken away, nature had taken over, and it was with great difficulty that we located the site of our house.

Chapter 7

It would be nice, after having served four years in the armed forces, to be able to look back to at least one occasion of direct confrontation with the enemy, but mine was a support role. The nearest that I got to anything approaching combat was when my company went ashore at Balikpapan on D+I. I almost said "stormed ashore" which would have been something of an overstatement. I guess the most exciting thing for me was wondering if the sea-proofing of our vehicles and heavy mechanical equipment would stand the test of rolling off the LSTs into four or five feet of water.

I had left Fiji early in September 1939 in a great surge of patriotic ardour to enlist. I should admit that patriotism to England, the land of my birth, was the real incentive. Australia was not directly threatened at that time other than by the prospect that Europe and Britain would be defeated following which Hitler would have continued his conquests and in due course brought Australia under his domination. It was a similar impulse that motivated thousands of others to enlist.

Following his devastating success in France, the feeling grew even more intense and the numbers volunteering increased. My first attempt in 1939 had failed because of my defective eyesight and on two subsequent occasions I was rejected because I was employed in gold mining which was a protected industry. Gold earned much needed hard currency. The situation changed when America entered the war following the Japanese attack on Pearl Harbor and, when the two Royal Navy battleships, the Prince of Wales and the Repulse were sunk off the coast of Malaya by Japanese Kamikaze attacks, I handed in my resignation and offered myself to the recruiting officer as unemployed. I had no difficulty in passing the eyesight test on that occasion as, profiting from my earlier experience, I learned the testing chart by rote in the few minutes before being told to take my glasses off and start reading.

Becoming a private soldier and relinquishing all the personal status that one possessed as a civilian is a very testing experience that takes quite a lot of getting used to. It comes as a series of shocks. The first shock comes with the initial outfitting. My size in underwear, uniform, socks and boots was roughly appraised by eye and the garments slapped down on the counter with a kitbag to take them away in. The second shock comes at the first wearing. Any ideas that a person might have about what a fine figure of man he is and how well he carried his clothes are immediately dashed. Being ordered around by some elderly relics of an earlier war, identifiable from the rest of the khaki clad figures by their general decrepitude and the stripes on their uniforms, also takes some getting used to, especially as members of this group derive a great deal of pleasure in the exercise of the authority which these same stripes give them. The final disillusion comes when mixing in public in the new garb. Instead of the wearer being treated as a potential saviour of the nation who has sacrificed all the comforts of home for the general good, he is treated as a stupid moron who must be devoid of brains and is incapable of doing anything better or he wouldn't have joined up.

Once in, what motivated me to get out of the rank and file was the shockingly poor quality of the NCOs and commissioned officers charged with the responsibility of teaching me how to become an efficient soldier. I became difficult to get on with and it wasn't long before I was categorised a trouble-maker. This is in no way my normal behaviour but I did find it necessary to organise a small group of people with like views to myself into a sit down strike. Our main cause of complaint was that we had been at the training camp for a couple of weeks and had received no training whatsoever wholly for the reason that there was no weaponry on the use of which we could be trained. I was paraded before the CO of the camp for discipline, an interesting occasion which resulted in my transference to another smaller camp at Narrogin where there was an engineers' unit. The officer in charge there, a captain, was either a retired or a renegade school teacher with whom it was possible to discuss things rationally, with the end result that I was recommended as suitable officer material and sent to an Engineers' Training School at Casula just outside Liverpool in New South Wales.

This trip to Sydney to the School of Military Engineering was notable not only for the length of time it took but for an astonishing event at its termination. On passing through Kalgoorlie I called on a friend of my brother's working as a teller in one of the banks there who told me that Frank, who had conscientious objections to serving in any capacity requiring him to take arms against anyone, was working on the merchant marine between Sydney and various ports in Papua New Guinea. If I wrote to him care of the GPO Sydney it was certain that during my thirteen week stint at the SME Frank would be in Sydney and we would be able to meet up.

The rail trip from Kalgoorlie to Port Pirie was interminable. The additional traffic over the line due to troop movements and army equipment had run the locomotives into the ground and this, combined with the frequent stops at messing points along the track to supply the troops with hot water to make tea, resulted in very slow progress. Food along the track consisted exclusively of bully beef, dry biscuits and apples, a couple of cases of each being deposited on the coach platform to be issued three times a day. I cannot recall how long it took to get to Port Pirie but the total trip including the Pirie-Adelaide-Melbourne-Sydney sections took two weeks.

Private Brodie-Hall WX-19456

It was a Saturday morning when I got out of the Leave and Transit Depot at Sydney and I went straight to the GPO in Martin Place to send Dorothy a telegram telling her of my safe arrival. It was just 9.00 a.m. and as I approached it from Martin Place I had a strong premonition that I would meet Frank there. At that hour the GPO was almost deserted and I walked head on into him in the Postal Hall which he had at that moment entered from Castlereagh Street. He had come down from Port Moresby, his boat had just docked and he made for the GPO to ring his girlfriend who worked at Grace Brothers, to tell her of his arrival and make arrangements for the weekend. We had a very enjoyable day together and then parted, for him to take his girl to a show and for me to check in at the SME out of Liverpool. Two days later, on Monday, Frank was posted to a ship bound for London. He had a shocking trip along the English Channel being heavily strafed from enemy aircraft and was very pleased to be able to spend several weeks at home with mother. His next posting was on the *SS Ceramic* which was torpedoed and sunk off the Azores in December 1942. There were no survivors. The meeting with Frank that Saturday morning, hatched out of the infinity of time and space, was the only chance I had to see him before his tragic death. If I ever needed convincing that there are external forces which influence our lives that meeting with Frank convinced me.

The course was interesting and not difficult for anyone with an engineering background who undertook it with due diligence. It lasted thirteen weeks and after graduating as a first-lieutenant I was posted to the 6th Field Company RAE, a militia unit formed in Ballarat and on station in WA. The personnel were mostly very young, in the nineteen to twenty-two age bracket, including the NCOs and were in the process of converting their militia status to regular army.

While I was strong on engineering I was totally lacking in any experience in the day-to-day running of an army unit. My sergeant, Alan Widmer, was young but experienced and I quickly came to terms with him that, until I learned from him the essentials, he would continue to run the unit from day-to-day. He was a very likeable fellow and we made an effective team. After a few months I was transferred out of the unit into a Boring Platoon and I lost touch with him until later in the war after the 6th Field Company was transferred into the 2nd Mechanical Equipment Company to which I was posted as 2IC with the rank of captain and we resumed a friendship that has endured until the present.

In the period between joining the 6th Field Company and being posted to the Boring Platoon we were busy instructing the young recruits in the skills and practices of military engineering. One of the camps we occupied was sufficiently close to my home at Gooseberry Hill for it to seem reasonable to me that I should spend my nights there in a comfortable bed rather than on the ground in a tent. The OC quite rightly prohibited my use of the utility which went with my rank but went further and disallowed my leaving camp until 7.00 p.m. after evening mess. It was nine miles from the camp to home, the last mile or so being up the side of the Darling Scarp. By alternately running and walking five hundred paces I could cover the distance in one and three quarter hours arriving home at about 9.00 p.m. I would leave again at a little after 4.00 a.m. arriving back in camp in time to take the platoon through their physical training which consisted of a mile jog around the roads. This became a nightly event for the five weeks we were camped there. There was one poor fellow, Sapper Swain, one of those chaps who don't start to function until about 10.00 a.m., and who on these occasions was always at the tail end of the platoon complaining bitterly that Australian Army Regulations strictly forbade a soldier to be subjected to such an exercise until he had been nourished with a cup of coffee and a biscuit. I refrained from telling him that I had just covered nine miles before taking the platoon out.

If it is thought that as a result of this exercise I must have been in tip top training there is an interesting follow on from this. When the company joined the brigade at Mandurah I had been out in the bush with my platoon and returned with them one day at about midday to learn that brigade sports had been arranged for that afternoon and that we were expected to field some competitors. I paraded my boys, broke the news to them and called for nominations for various events in which I knew we had some likely colts. I got a nil response, no volunteers, so I told them I knew of their individual capabilities and if they didn't offer voluntarily I would detail them to compete. After nominating a couple of chaps for the one hundred yards and high jump, someone said, "It's OK for you to put us in but what about you, sir?" "Right," I said, "I'll run the mile." After that nominations came in more freely.

The track around which the races were contested was on a depressed area upon which there was a strong growth of samphire, a dense resilient plant which was very spongy to walk on. Competitors could run either in bare feet or in their army boots. There was no alternative. The soles of my feet are very sensitive and I elected to run in my boots which, as a point of interest, weighed seven pounds. I do have rather large feet. The time for the mile event came around, there was a large number of entries and at the crack of the starting pistol, my sergeant Alan Widmer and I took the lead with the largely barefooted throng following. We were still in the lead at the end of the half mile and as we entered the third lap I heard one of my sappers say, "Look at the poor old bastard, he's going to fall arse over head in a minute." They were prophetic words. I hung on for another forty or fifty paces and then collapsed in a heap. A little chap about half my size came out of the ruck in the last quarter mile and sprinted home to win by about a hundred yards. It was incidents like that that helped build up a very good rapport between me and the lads in the platoon.

☆ ☆ ☆

The surges of patriotism I experienced when I offered myself on several occasions to the recruiting officer didn't last long after it became apparent that my service was most likely to be spent in Australia. It would be less than honest if I failed to mention that the let-down at this prospect was mitigated by recognition that serving on the mainland was safer than being in the forward areas in the Pacific arena to which the centre of gravity had moved insofar as Australia was concerned. For the next couple of years it was rather like doing a civilian job in military clothing. I was fortunate being in an engineers' unit as we were employed for nearly all of the time. Even then it was difficult to keep our troops from grizzling. It must have been very difficult to maintain morale in the infantry units. One of the jobs I disliked most was to be made o/c train on those all too rare occasions when I was travelling to and from home on leave. It seemed that the troops on any train to which I had the misfortune to be posted as o/c were made up of small detachments from many units travelling without any of their own officers. These are circumstances in which even the mildest private, if carrying sufficient liquor internally, tends to become insubordinate and uncooperative. Part of the o/c train duties was to check each individual's regimental details against a master roll. Trying to complete this task when most of them were either drunk, in a drunken stupor or ready to take on all comers was quite a job. However I had two experiences in troop train travel that had elements of humour.

On one occasion travelling back on home leave I contracted mumps. We had travelled all night in suburban rail cars from Melbourne to Adelaide when, just as day was dawning and the first rays of light filtered in to the carriage, the chap opposite me burst into great guffaws of laughter and announced, "You look just like a cow with lumpy jaw". He must have been a farmer! "You've got the mumps." I always thought it was a kids' complaint but his diagnosis was correct and I finished up in hospital in Adelaide flat on my back in bed. After ten days or so of this I had enough and persuaded the staff to let me out to try to get on the next troop train going west. It was a bit of a battle as the Port Pirie-Kalgoorlie train was choc-a-bloc but I was allotted a position in a compartment in a coach reserved for officers only to find that it was fully booked. The normal two berth coach accommodated three, one in each upper and lower berth and one on the floor. These had already been allocated, according to rank among the other officers, of whom one was a navy commander, another an army major, the other an RAAF flight lieutenant. The latter, being the lowest ranking of the three, had the floor. I as an army lieutenant was very much supernumerary and as my gear was taking up space that the flight lieutenant wanted he went down to the guard to complain. In a short time the guard appeared to inquire "which is the gentleman that hasn't got a berth?" "Him", they said in unison pointing accusatory fingers in my direction.

"Alright", said the guard. "You come with me and I'll find a place in the van for you to stretch out in." In thanking him for his kindness I told him I had just that morning got up after ten days in bed and felt a bit shaky. He agreed that I did look a bit peaky and showed me into a compartment in his van in which a berth was made up with white sheets and pillow cases. "You get in there, sir and I'll bring you a nice hot cup of tea." I don't know if it was his own berth or whether it was one he kept for special guests. During conversation it came out that his people owned the Oriental Hotel in Kalgoorlie. We had quite a long chat about the old town. Next morning the flight lieutenant came down to see how I was getting on and found me propped up on my lovely white pillows eating bacon and eggs with toast and drinking steaming hot coffee. He didn't have much to say. In fact I think he was speechless. I was quite sorry when we had to change trains at Kalgoorlie where there was a break of gauge from standard gauge to narrow gauge and I lost my comfortable berth and personalised service and reverted to sleeping in my uniform with the rest of them.

The other occasion had rather a more sombre note but although it made us hopping mad at the time was really quite hilarious. I and my fellow officers were entraining our company at Adelaide Central Station. We had prepared for a rather long comfortless trip by pooling our money and buying some tasty rations for the way. There were couple of seasoned baked rabbits, a couple of roast fowls, fresh bread rolls, butter, cheese, pickled onions, gherkins, olives and anything else you can think of. The railway station was rather congested. There were other units than ours to be loaded and it became very much a matter of defending your own territory while making sure your troops didn't invade someone else's.

In the general melee the goodies became a damned nuisance and to protect them while the rest of us were attending to our train loading duties I asked a dear little old lady if she would stand guard over them for a few minutes. She was just the sweetest old thing, any young man, or woman for that matter, would have been proud to claim her for his grandmother. Petite and quietly dressed, white curly hair tucked discretely under a neat hat, eyes softly gleaming behind gold rimmed glasses perched on a cherubic face, she assured me she would love to take care of our parcels. I should have made sure that she was going to take care of them for "us". Five minutes later with every one safely and correctly entrained this sweet little thing and our parcels had disappeared, vamoosed, vanished, evaporated into thin air. Before we could mount a hue and cry our train gave a toot and departed and that little old lady and our provender were seen no more but remained a topic of conversation between us for several weeks.

I was pleased when I was transferred to the Borers as during the time I had been with the Field Company we had been playing war games, getting accustomed to the standard army bridging equipment, folding boats, pontoons, small box girders, and practising throwing together bridges out of rough timber locally procured. This was the time when Australian troops were being hurriedly withdrawn from the Middle East theatre to repel the southward advance of the Japanese. While many of the fighting units were posted directly to New Guinea, many of the service units were sent to Bonegilla in NSW for reformation and re-equipment. The 6th Field Company RAE by then had voluntarily relinquished its militia status and converted to AIF. Most of the personnel were transferred to the 2/2nd Railway Construction Company, but I suppose because of my mining association I was posted to the 2nd/2 Boring Platoon, a small outfit about seventy strong whose function was to develop ground water supplies. This entailed selecting sites favourable to the recovery of water and drilling the required number of wells to establish an adequate supply. The Railway Company, while it may have had a useful function as such in the Middle East, would have no place in the island fighting of the Pacific war and, on being reformed, changed its name and function to road and air field construction and other similar work and became a Mechanical Equipment Company.

When I was transferred to the Borers, it took me to the Darwin area where we were given the task of drilling sufficient wells to maintain the presence of two army corps. Apparently at that time the Northern Territory was the preferred place from which to launch the assault on the Japanese territorial gains in South East Asia and the Pacific. Great urgency was given to our project which we progressed on a three shift programme seven days a week. The platoon was manned to allow for this. That there was no break for recreation was no sweat as there was no where to go and nothing to do anyway. We were working eight machines, percussion rigs, commonly referred to as mud punchers and the crews consisted of a corporal in charge, three drill operators, a blacksmith to sharpen the drill bits, a mechanic, a transport driver and a cook. All hands weighed in when the machine was moved to a new site and camp was struck and when the hole was cased in difficult ground. Otherwise the only chap visible doing any work was the drill operator. The rest of the crew could usually be found either sleeping, reading, playing cards, out shooting if the weather conditions were favourable, or as it was a mineralised area, prospecting. This situation was a cause of great distress to the Chief Engineer (RAE), Colonel Oscar Myer, who would drop in from time to time to carry out an impromptu inspection. He couldn't understand that if only one man was working and the rest were languidly lying around that everything was going along very satisfactorily. He wanted everyone to be up and doing. If ever we had prior news of an impending visit we would do our best to stage something for him. I think it made him happy but it did nothing to increase the footage drilled which was my main concern. The unit drilled one hundred and eighty holes all cased and capped pending the arrival of the troops that never came. The boys in the Boring Platoon were justifiably annoyed when an order came through to pull the casing, which resulted in the collapse of the wells, and to quit the area. At one stage during the work programme the urgency was such that their leave had been cancelled, an action that made them very sore as on their return to Australia from the Middle East they had been transferred directly to the Northern Territory without any home leave. Queensland had become the preferred jumping-off place for the island assault.

In so far as it is possible to enjoy such an unproductive period I enjoyed my time with the Borers. They had been largely recruited from the mining industry and quite a few came from Kalgoorlie. They had minimal respect for authority beyond that inherent in their own officers and on many occasions I had to get them out of serious trouble. I cannot remember whether the Boring Platoon was disbanded. I was transferred to the 2/2nd Mechanical Equipment Company with the rank of captain and the standing of 2IC and very pleased to find a large number of the old 6th Field Company among the ranks including my sergeant, Alan Widmer. The interim camp at which the company was brought up to strength both in men and equipment was at Petrie, a camp a few miles north of Brisbane and it was there I met my new CO. He was a very languid chap who didn't like work in any shape or form. This suited me quite well as I was able to assume the greater responsibility in running the unit than would have been possible under ordinary circumstances. He was a most elegant calligraphist and I was able to keep him happy without overtiring him by giving him printing assignments.

The nature of the services the unit performed required small groups of men and equipment to be detached very frequently and at short notice for varying periods and, as the urgency of the individual assignments varied widely, it was essential that the whereabouts of men and equipment could be identified at a moment's notice. To achieve this in the field I had a number of shallow, readily transportable cabinets made, which accommodated a large number of miniature trays, fabricated from light gauge galvanised metal to contain the individual personal details of each member of the unit – name, number, blood group, rank, trade grouping etc. – as a permanent record and also the transient information of his movements by location and duration. Each item of equipment was treated in the same manner in other cabinets. The system worked like a charm. Initially the name of the person or item of equipment was displayed on the end of the tray written in pencil on an adhesive label but later this became the ideal medium for the exhibition of the major's artistic skills. Replacing the originals with most beautiful hand-crafted lettering in different colours to facilitate identification became his main occupation for many weeks and kept him feeling that he was playing a significant military role.

He was a very easy fellow to work with, contented with the exercise of his hobby and the pleasure of handing out summary justice at the defaulters' parade. His officers ragged him constantly but surreptitiously and took special delight in playing on his nerves and persuading him on occasion that he was sickening from some undiagnosed complaint. This was usually accomplished at breakfast before he had decided that it was alright to speak to anyone and was still conversing in grunts. An opening remark, sotto voce, but just loud enough for him to hear, to the effect that, "The major doesn't look too well this morning," was later followed by solicitous enquiries of him of his health with increasing accent on how poorly he looked. This would soon have the poor fellow admitting that, "Yes, he was feeling rather unwell", and his speedy retirement to his tent and bed where he would stay for two or three days until he felt well again. The most blatant exhibition of this weakness occurred when we embarked from Morotai, to which we had been moved in the process of the invasion force being marshalled for the assault on Balikpapan.

During our brief stay at Morotai, the CRE, a very hands-on Queenslander, the complete antithesis of our CO, had been putting the finishing touches to our post-landing duties and made it quite clear that bombs disposal, in which all the company's officers had been thoroughly trained, was the responsibility of the officers alone and would NOT be delegated to other ranks. That was enough to send the CO to bed a day before we embarked and to keep him there for the best part of a week. He rejoined the unit when everything was settled down and his batman had established his tent, complete with floor covering, to the standard to which he had become accustomed.

I have good reason to remember, but with humiliation, Balikpapan where we landed on 4th July to the accompaniment of a first class American Independence Day show provided by several units of the US navy anchored at a favourable distance to permit some good target practice to accompany the celebrations. When the roads had been cleared of mines, repaired and made trafficable, the next urgent task was to prepare a site for a hospital. The site selected, nicely elevated with a view to the sea, was pure sand which required sheeting with gravel and consolidating to provide a firm weatherproof surface for both the field hospital site and the parking apron. The unit had two, if not three road engineers from various state Departments of Main Roads (DMR) around Australia, who immediately set off to reconnoitre the environs for gravel but who not only were unable to find any but were firmly convinced that there was none to be found and no sense in looking further. I was indignant at their lack of initiative and resignation to the situation as they saw it. I made my own search and found a supply of what I believed would do the job. When the field test to establish whether the clay content was satisfactory was applied, the material proved to be marginal but as there was none other available nor any ground reinforcing steel mesh to be had, it all being required for air field construction, I gave orders to use the stuff I had found.

Great progress was made in excavating, loading, carting and spreading it and it looked a first class job. My contempt for those DMR trained officers who had given in so easily appeared to be fully justified. And then the rain set in. It rained with great regularity every afternoon for several hours commencing at about four o'clock, great bucketing downpours which progressively reduced the gravel to the consistency of porridge. Daily efforts to incorporate some of the underlying sand with it to improve its draining qualities failed miserably, nor did it impart any stability to the sand. While this fight against nature was at its height I had a visit from the adjutant of the chief engineer's, HQ. He also was a road engineer and the scale of the problem I had created caused him to express the pious hope that I knew what I was doing. I knew what I was trying to do but by this time had lost all confidence in the prospect of success unless it stopped raining immediately and never rained again. The next morning he rang to say that he had located some air field reinforcing mesh and to get rid of my gravel the best way I could. It was an elevated site and this was quickly achieved by pushing the muck over the edge. I still cringe when I think of this and ponder the consequences if the fighting had gone against the allied forces and casualties had been heavy and the need to use the hospital acute. The event remained a source of amused discussion with the road engineers for the rest of my association with them.

I wasn't the only one to cast aspersions on their knowledge and experience. A high ranking officer, I think he was a brigadier, visiting Balikpapan decided that he would like to play tennis and demanded that a nearby court that had suffered damage from the shellfire should be repaired and ready to play on a couple of days later. One of our chaps, I think he was the one from the DMR New South Wales, inspected the court, looked over the bituminous materials available to effect the repair and stated unequivocally that it would not be fit to play on as early as the general required. Perhaps there was an additive which we did not have which would have speeded up the setting time of the bitumen. Our man assured us that in the Borneo climate the stuff he used would take at least a week to set and not the two days he had available. When this negative piece of news was conveyed to the brigadier, his reply was terse. "You are engineers, aren't you? Get it fixed." When the appointed hour came, it was hardly possible to detect where the damage had been, so carefully had the holes been filled in with bitumen that they were barely discernible. The game had scarcely started when the general, positioning himself to receive a service, stepped into one of these booby traps which immediately seized his shoe. His forward momentum pulled it off his foot which came down on another patch and neatly removed his sock, leaving him precariously poised on one leg and not knowing where to stand next. Our chap, who had been watching the event with great interest, beat a hasty retreat, arriving back at camp just in time to receive a terrific blast from the brigadier himself.

The Balikpapan campaign was short-lived, terminated by the capitulation of the Japanese on 15th August, 1945 following the atomic bomb attack on Hiroshima and Nagasaki. Word of the war's end came through as a rumour tied in with the use of an atomic bomb, something of which no-one had previously heard and which sounded most unlikely. Official confirmation reached us after dark and was celebrated by much unauthorised firing of rifles. As there was no champagne to be uncorked, or beer either for that matter, to celebrate the occasion that seemed to the troops to be as good a substitute as any. A couple of the lads took the opportunity to put a couple of rounds through major's tent. They were placed fairly high up on the likely assumption that he would be lying down anyway and out of harm's way. Like Queen Victoria, he was not amused. As he didn't know the culprits, he couldn't do much about it and wisely decided to take it as a good joke.

After the end of the Pacific war the troops could see no reason why they should not be immediately returned to Australia and demobilised. It was something of a problem to keep them smiling and happy. To some extent I alleviated their boredom by digesting books by speed reading them and then presenting them to the troops in abbreviated form. I recall the most success was achieved with a review of a Penguin book called, I think, the *Tennessee Valley Authority* by David Lilienthal. I have a good recall for figures and there were sufficient statistics in this book to make my verbal abridged version interesting and it commanded a high degree of respect.

Musical afternoons were also provided, when we won over a quite sizeable proportion of the chaps from jazz to classics. We won a decisive victory over one chap, a transporter driver who, replying to my query as to how he liked the classical items which we interspersed with the popular tunes, said, "Oh, they're alright I suppose but there's not much tune to them; you can't whistle them." At the next session we opened up with Mozart's "Eine Kleine Nachtmusick" and then played snatches of it throughout the afternoon between record changes. On getting his assessment after that session he admitted that, "Yes, there may be something in it, there was one tune . . . " On being pressed to identify it and playing dumb to his efforts to verbally describe it, he was forced into whistling it, much to his surprise. "By cripes, you can whistle it." The battle was won and before we lost him back to civilian life we had educated him in the delights of a number of classical composers.

As all of the other officers in the company had served in the Middle East campaigns before their Australian or Pacific area service, they were the first to be demobilised, leaving just the major and me to wind the unit up. There was a reasonable sharing of the work until we arrived back in Australia and were camped this time on the outskirts of Brisbane. I must admit he was the devil with the women and within a couple of days had found himself a girlfriend of similar age, forty-five to fifty, who owned either a dress boutique or a beauty salon in Brisbane. His normal routine from then on until he was discharged was to leave camp at one or two o'clock in the afternoon and not return until 7.00 a.m., just in time for breakfast. Before this he had always been a pernicketty eater but the change of activity must have done him the world of good because he developed a voracious appetite and knocked off a breakfast fit for a navvy before excusing himself and retiring to bed for a few hours. He once confided in me that I was a great chap but, "Those other fellows were a lazy lot." Eventually the major got his discharge and I was left with final details, concluded at a camp in Victoria, Broadmeadows perhaps, and I too was free after almost four years of unproductive effort but heartened by the knowledge that I had responded when I was needed.

Chapter 8

For some time while I had been under-employed in the military forces I had been taken with the idea of starting afresh in the field of hydroponics, a developing technique which had proved useful in providing on the spot vegetables at places in the Pacific like Guam. I had acquired a couple of books and the technique seemed simple enough for anyone with a smattering of chemistry and a liking for growing things. It had seemed like a great idea and had been the subject of many letters home. When I got back to the west and faced the reality of providing for a wife and two children, it showed up for what a nice romantic and impractical dream it was and I realised the sooner I got back into the mining industry the better.

I had completed the first two years of a four year Associate Diploma course in metallurgy and the sensible thing would be to complete it. At my age, I was then rising thirty-six, unless I went into it head first, it would prove to be a slow protracted business. Fortunately the Commonwealth Government had foreseen the need to provide assistance for those who had interrupted studies to join the forces to have the opportunity and the means to finish them.

The system that was set up under the title of the Commonwealth Rehabilitation Training Scheme was intended for twenty year olds who fell into this category and I was very much outside the age bracket. As luck would have it, a senior official administering the scheme in Perth, Dr Bertie Moore, was a former principal of the WA School of Mines in Kalgoorlie. My scholastic record was pretty good. He also was a close golfing friend of my father-in-law and with him putting in a good word for me I was granted an interview and accepted.

I believed that with a little co-operation from the staff of the school, from whom I needed approval to take the remaining two years of my course in one year, I could get through. The conditions precedent to success as I saw it were that I would have to leave my wife and family in Gooseberry Hill for the duration and live with her parents in Kalgoorlie. I thought it not possible to engage in an exacting course of study and have the distraction of a young and active family. Dorothy agreed without demur although we had already been apart for four years. I am sure that she recognised it was the only way if we were to improve our lot.

The CRTS paid something towards the cost of living. I have always been a good saver when the occasion demanded and my accumulated savings plus the deferred pay allowance amounted to about £1250. Against today's highly inflated currency this will not seem to be very much but viewed against a background basic wage of about £4 a week and compared with today's minimum wage of $400, it would have been equivalent to about $60,000 in today's terms. It appeared that we had enough money to sustain me through the course and have a sufficient residue to keep us going until I found satisfactory employment. I cannot remember if I made much of a thing about leaving wife and family again for the best part of another year. I am sure that I would not have liked it but by this time I was hyped up with the prospect of resuming my studies and finishing the course in record time. I had a little difficulty in persuading the lecturing staff that I should be allowed to compress the two years into one. This was understandable as some of the subjects were introductory to others and they believed that they should be taken in sequence.

My age helped me in these discussions. Academics tend to live in a different world to ordinary folk. I had a firm resolve and eventually they caved in, tempering their agreement to this most unorthodox behaviour with dire threats should I fail and wish to continue under the scheme in the following year! There was the additional handicap in that the first term, it was terms in those days not semesters, was half way through and incurred a lot of catching up. When the results of the first term exams came through I knew that if I could keep it up that I would be successful. It was fortunate that my friend Denis Martin who had accompanied me to Fiji had signed up under the scheme also and was studying a course in Mechanical Engineering. He was married and he and his wife Norma lived in a small flat, a converted shop I think, in Maritana Street not far from the school. The friendship between Denis and me lasted over several decades until his death in the late 1980s, after he had established and developed the business of Warman International, which amongst other things is one of the world's most successful manufacturers of pumps for the mining and dredging industries.

My in-laws lived about two miles from the school and walking there and back three times a day kept me in very good condition. Every third weekend I could go down to Perth by train, returning on the Sunday evening train to arrive back on Monday.

The average duration of the train trip was about twenty-two hours. Shortly before the war Hubert Opperman, Australia's world famous cyclist, engaged in a race, Kalgoorlie to Perth, with the Kalgoorlie Express. All very exciting. The express stood drawing deep breaths at the station and puffing out thin wisps of steam, obviously conserving energy, Oppy in the parking area outside with his leg cocked ready to throw over the saddle. At last the guard gave the green light, the engine whistled threateningly and emitted a huge cloud of smoke as the driver admitted steam into the pistons. Oppy mounted, his back wheel throwing up a cloud of goldfields dust, and the race was on. There was an eight foot strip of bitumen from Kalgoorlie to Coolgardie, after that it was gravel all the way to Northam and not just smooth gravel but heavily corrugated road for nearly all the way. Oppy covered the distance of three hundred and seventy-five miles in record time, arriving several hours ahead of the train, A truly epic race. However, even if the train did lose the race, the rail service wasn't all that bad, for a small fee, five shillings I think, a sleeping berth could be had and the guard would wake his sleeping berth passengers at 7.00 a.m. with a cup of tea. The train arrived at 9.00 a.m. in good time for a day's business.

These weekends were very welcome. The children were growing up and I was missing so much of their childhood and although Dorothy was a strong and very capable woman who had no qualms about cutting the wood for the fire, I had the old-fashioned idea that that was man's work as was digging the garden and similar manual chores.

Towards the end of the year a favourable circumstance arose which was one of the most significant events in my life. My in-laws were part owners of a tennis court on which they and a group of elderly mining company executives played most Saturday afternoons during the year. Whenever Lindesay Clark, the managing director of Western Mining Corporation, visited Kalgoorlie he was included with those playing. One Saturday afternoon in the winter of 1946 I was working stolidly away in the dining room of my in-laws' house and feeling a bit jaded when it occurred to me that it was just about the time when afternoon tea would be served at the tennis court and that I could do myself a good turn by walking over and joining in. I knew all of the players except on this occasion Lindesay Clark, to whom my mother-in-law quickly introduced me, sketching in my general background and what I was currently engaged on. Mr Clark had just finished his set and elected to sit the next one out and to talk to me. Well, that is hardly the way to describe it; to delve unobtrusively into my past activities, present occupation and future aspirations. I think that

I caught his interest because of two things. He also had interrupted his education when he enlisted for the 1914-1918 Great War and had resumed his studies at Melbourne University after demobilisation. He was then in his early twenties. He was very interested to learn that I had done the course at the School for Miners he had introduced at the Triton Mine. He didn't play any more tennis that afternoon and I didn't study any more for he kept me answering questions and discussing varied topics with him for the next couple of hours during which time he proffered several pieces of advice.

Having ascertained that I had as much experience in mining practice as in ore treatment procedures he recommended that I should immediately drop the metallurgy course and take on the mining course. "Metallurgists never get the top job, that's for the mining engineer", a statement which for those days was entirely correct. I jibbed at that. I had already had to pick up half a term and didn't like the idea of starting half a year behind by reverting to mining. He had however aroused my interest and I told him that I would seriously consider continuing on for a second year and knocking off the residual subjects in the mining associateship. The two courses had about half of the subjects in common and taking on the mining course the following year would have again entailed compressing two years work into one. There was still just sufficient money left in the kitty to carry me through the following year and I was considerably encouraged by Mr Clark's remark that when I had successfully completed my studies I should consult him regarding future employment.

At the end of the first year when I had passed all of the subjects with very good marks and had only the special assignment (I don't think it was dignified by the title of thesis) to complete, I used as my subject an analysis of the treatment process designed for the Beryl Corporation. Dorothy agreed that I should continue on during 1947 and qualify as a mining engineer. These were brave decisions on her part expressing a high level of confidence in my ability to bring home the bacon. It was an interesting time. I was twelve or fifteen years older than most of the other students. When I enrolled for the second year it prompted the comment from one of the locals, not a student, "That I was a fugitive from economic reality, dodging through protracted study the need to earn an honest living!" or words to that effect.

The year was running out when in October 1947 I heard that Lindesay Clark was in town again and it seemed that the time was ripe to remind him of his advice which I did in a short letter drawing to his attention that I had followed his recommendation and would complete the mining course that year and was now applying for a position with Western Mining Corporation Ltd. Within a few days I was being interviewed by him in the presence of Frank Espie (Senior) who intimidated me greatly by sitting with his arms tightly folded over his chest and submitting me to a cold unremitting stare for the duration of the interview. I am sure that if he had known the dampening effect that his presence had on my confidence he would have withdrawn. I came to know him very well until his death in the mid sixties and I have never known a more gentle man or had a more sincere friend.

The interview was interesting in that the way it was conducted and the questions asked indicated quite clearly that the focus was on whether I would make the top echelon or not. When on being asked in what area I would like my first appointment to be made (assuming we got that far), I replied in geology. That evoked the encouraging comment that he was pleased about that as the future policy of the corporation would be to encourage senior mining operators to have a sound knowledge of geology. He elicited from me that my main requirement in taking employment with whoever, was scope for advancement within the organisation. His answer to that is firmly embedded in my data bank and will respond with instant recall. “I will put you in the milieu in which opportunity exists, after that it will be up to you.”

When it came to starting salary I admitted that my monetary expectations were not great as I had been out of the industry for six years, due to military service and the two years of qualifying study. My main requirement was opportunity for promotion and, as I had been apart from my family for six years, a house was a must. He agreed that he could and would satisfy these conditions pursuant to me successfully completing the course, a point upon which I assured him there was no doubt. It would not have taken a great deal of effort to complete, in due course, the associateship in geology, and in anticipation that this would come about I aimed to complete the assignment, the subject for which I had decided would be the detailed geological mapping of one square mile around Mount Hunt, a prominent feature south of Kalgoorlie named for the pioneer surveyor who in 1860 had surveyed much of the country east of Southern Cross and some distance east of Kalgoorlie.

Most of the surveying and mapping was done at the weekends although I used one inter-term vacation to get a fair slice of it done. I needed someone to carry the staff and on occasion the chain, although much of the work was done using a plane table. In the first instance I had the services of a Canadian, a man of about fifty years who I had known when we were both lumping wheat and who was at the time employed as a miner on the Great Boulder. He was a single man and pleased for the opportunity to fill in the weekend in the open air. We made a habit of camping out on Saturday nights to make the best use of the time. The weather was quite fresh and we were a bit short on blankets and sleeping on folding canvas stretchers which let the cold breeze through so that usually we were both awake by two or three o'clock – at least that is what Harry pronounced it to be. At some time in his career he had been a sailor and knew the stars very well and would make such statements as, "Going on Achernar's position it's about three o'clock". Sometimes he would refer to Canopus or Aldebaran. As neither of us possessed a watch there was no means of checking on the accuracy of his guesses nor did they do anything to shorten the night.

I have an idea that the law caught up with him or maybe a wife that he hadn't told me about for suddenly he stopped coming and I did not see him again. But all was not lost. A young fellow, Alan Watson, one of the younger members of a Boulder family who was studying at the School of Mines, volunteered to be my helper and I was able to finish the job. I lost track of him for many years. He left Kalgoorlie and finished a mechanical engineering degree in Melbourne. When I next met him he was general manager of the construction of the Melbourne Underground Loop Line which was very close to completion and over which he gave me a comprehensive tour. I experienced a great glow of pride in his achievement.

It had been a tiring year and I took a month's holiday during December 1947 and when the offer of a job with Western Mining came through, I reported for duty on 5th January 1948. WMC was at an interesting phase of its development. The directors had adopted a strategy that instead of focusing on single mines, the target would be to develop districts that would include multiple mines on the theory that if three or four or five mines were being worked simultaneously and one or more were either being developed or had run out of ore, the others would maintain production at a profitable level until replacement reserves were discovered either in new deposits or as extensions of those from which production had temporarily ceased. The district within which this strategy was to be employed was the Yilgarn, specifically a section of the greenstone belt extending from Bullfinch in the north to Southern Cross, a distance of about twenty-four miles. The anchor mine upon which the largest reserve was known and at which drilling had indicated would continue at depth was the Copperhead mine.

When I reported for duty, work was concentrated on the compilation of ore reserve tonnage and grade for the Yilgarn deposits. The strata hosting the Copperhead orebodies at Bullfinch are extremely contorted but the ore blocks were generally fairly wide and amenable to open stoping and the grade estimated at seven and a half grams/tonne would be attractive should the gold price improve by ten per cent or fifteen per cent. As the directors were confident that the price would rise in the near future they felt well justified in advancing preparations to bring their plans to fruition immediately the cost/price ratio was favourable. It was interesting to be involved in that phase of the work even if only for a couple of months until I was transferred to Central Norseman as junior mine geologist subordinate to the senior geologist who was responsible for mine and district exploration and the supervision of my work which consisted of mapping underground development headings and keeping the miners on track, supervising the stope miners, the underground diamond drillers and the samplers on both the Mararoa reef and the Princess Royal reef.

The prospect was exciting and I presented myself at the mine full of enthusiasm which was very quickly dashed when I learned that no-one knew anything about my new appointment or indeed of my existence. A telephone call to Kalgoorlie soon put that right but the mine manager knew nothing about providing a house in the staff precinct for me and there was none available. A house was found for me in the "compound", a cluster of twenty-four small bungalows built some years before at an all up cost of £195 to house miners, with very small families I would think. They were very basic. Six hundred square feet inside the walls divided into four rooms with a three foot six inch passageway from front to back. Room sizes were about ten feet by twelve feet. There was an open front verandah and an enclosed back verandah which contained a bathroom at one end and a wash house on the other. The toilet was at the back fence.

The house allotted to me hadn't been lived in since it was built but the local children had obviously had a lot of fun in it. I swallowed hard when I broke the news to Dorothy who had to buy some essential furniture with the small amount of money she had been able to save during my two months employment with WMC, our nest egg just having lasted out my two years of study. Perhaps it was an advantage that the house was small and the amount of furniture required correspondingly small. It cost just £20 to furnish the house and pay the rail fare of the family to Kalgoorlie (home and mother for Dorothy). The furniture bought from John A. Bell, auctioneers, which consisted of a kitchenette,

House at Norseman

kitchen table and four chairs, a double bed and mattress, a very handsome large oak wardrobe, two beds for the children and a well worn sofa and two arm chairs, arrived before the family and I had time to arrange the items around this small house to create the best impression of a home that I could with the material available. The thought of settling down in a house for an indeterminate period had me standing on my tiptoes with excitement and they were within a day of leaving Kalgoorlie by rail for Norseman when Kalgoorlie had an unprecedented fall of rain – twelve inches in twenty-four hours.

There is practically no topographical relief and no defined drainage pattern in the eastern goldfields; an occasional isolated hill may rise two hundred to three hundred feet above the plain, the rest of the country is virtually flat. Besides flooding all the shops in the main street and the power station, it disrupted the rail service to Norseman for three weeks. Although I seethed with impatience, it gave me the opportunity to paint the kitchen furniture and establish an instant lawn. I had observed Adrian Brevett, the engineer from the adjoining mine, starting to dig out the lawn in front of his house because, as he put it, it was too much "bloody hard work to keep it cut". I persuaded him to hold his hand until the weekend when I would dig it out for him. By then I had become friendly with Peter Rafferty, the mine accountant, who borrowed a truck from the mine and helped me dig it out and relay it. Lifting it in slabs two feet square and six inches thick was heavy work but by Sunday night I had half the backyard covered. As one of the neighbouring miners complained, "You bloody staff men get well looked after. We have to do everything for ourselves."

Norseman was an unusual community, at least I thought it so, in that there was very little social spirit among the employees and, in my opinion, social cohesion should be generated from the top and should extend at the very least through the staff structure. It caused me some amusement and a lot of satisfaction when my mean little abode tucked away in the miners' compound became a hub, around which some staff social activities began to develop.

At that time Central Norseman NL was going through one of the cyclical downturns which has characterised its operations. The ore from one orebody was nearly exhausted before the next one, although indicated, had been developed to the point where ore production could commence. A part of my job was to re-sample the pillars that had been left for stope support and recommend those that could be profitably and safely extracted. These were very tense times for management and happily the situation was shortly after relieved by successful exploration on the Princess Royal reef when a diamond drill traverse intersected wide high grade quartz which developed into a very important orebody.

After I had been in the position of mine geologist for about eighteen months on a salary of £10 a week and no indication that there were plans to move me onto a more responsible position, I began to cast around for alternative employment, even to the point of applying for a job as plant superintendent at a mine in West Africa. I was waiting for an answer to my application when one morning, while I was underground at the Princess Royal mine, I had an excited call from Des O'Driscoll, the senior geologist, to tell me that Lindesay Clark wished to see me and to get back to the Phoenix mine as quickly as possible. When I arrived back and reported to Des, he was most excited and spilt the beans to the extent that he told me that Mr Clark was going to offer me a job and would I like to change out of my underground clothes, presumably to make a more favourable impression. I declined that and before long the great man appeared and after enquiring about some of the development work that I was supervising asked me outside for a chat. We walked up and down and around the diamond drill core farm for about twenty minutes or so while he quizzed me on this and that and ascertained that I was lecturing at the Norseman branch of the School of Mines, an activity that brought me in £8 a week, almost as much as my mine salary. Then he terminated the interview by saying, "Well, I haven't got anything for you. I just wanted you to know that you haven't been forgotten."

Des was all agog to know all about the job I was offered. I have seldom seen anyone as deflated when I told him the outcome. However, about two months later, on Mr Clark's next visit he offered me the position of Technical Assistant to the Managing Director, i.e. himself, in Melbourne office at a salary at least equal to my combined earnings as lecturer and mine geologist. He seemed to approve of the idea that I should refer the offer to my wife who, as I explained to him, owned a fifty per cent equity in me. I assume that he thought that any young matron offered the opportunity of changing residences from Norseman to Melbourne would jump at the chance and of course he was correct. Dorothy was delighted.

I was to leave Norseman forthwith and spend some time in Kalgoorlie office assisting Frank Espie Snr, general superintendent of Western Mining and to acquire some knowledge of the Corporation's other activities. This was a useful period mainly because I was instructed in project estimating by detailed synthesis of the proposed operation. At that time capital and operating costs were being estimated for the Yilgarn project. The price of gold in Australian currency had appreciated substantially following the devaluation of sterling in 1949 and the devaluation of the Australian pound pari passu with sterling, and a new company, Great Western Consolidated NL, had been floated to bring the leases into production.

While waiting for a call to proceed to my new appointment at head office, Big Bell Mines in the Murchison were urgently seeking the services of an underground shift boss to fill a vacancy until the examination for shift bosses to be held two months or so later would provide them with two or three from their own ranks. The Big Bell orebody was a model orebody for open stoping being six or seven hundred feet long and ninety to one hundred feet wide. Frank Espie decided that I should be loaned to Big Bell, as the experience I would gain would be of great value when planning the extraction method to be applied to the Copperhead orebodies. By this time I had been at Kalgoorlie office for five or six weeks living a bachelor existence and I had just been provided with a house. Dorothy arrived with the children on the Monday and I was despatched on Wednesday. I recall that Dorothy was not very happy about it but was wise enough to realise that it was all for the best. I didn't like it much either.

It was an interesting couple of months during which I got valuable first hand experience in sub-level stoping. When the results of the shift bosses' examination came through and the mine had been relieved of its problem, I wasted no time in returning to Kalgoorlie. My wife and I celebrated notification of our move to Melbourne shortly after by holding a small party for a few friends. It stays in my memory because the quarters we were living in were very basic and possessed neither refrigerator nor ice box and although it was early in December and very hot weather we successfully cooled the drinks under wet bags with an electric fan blowing on them!

I had to precede the rest of the family to Melbourne as I had to find a house to live in, which the company bought for me to honour Mr Clark's promise that I would not be worse off financially than in my previous job. On my first day he made what proved to be a very rash statement when he said, "You can settle in here for at least ten years."

One of my first tasks was to assist in the preparation of a brochure describing the operations that WMC was managing and to forecast the financial returns that could be expected. I wrote the first draft which I passed on to Mr Clark which he altered out of all recognition. This was at a time when W. S. Robinson was in

Dorothy and I shopping 1954. Candid camera in Forest Place. Everyone who walked through the precinct between 1940-1960 must have a similar picture.

Melbourne for some time and the amended draft was passed on to him. He was merciless with a red pencil slashing paragraphs and sentences out of the text, adding whole pages. He had been a journalist in his early life and he really had a ball with this draft. Eventually Mr Clark returned it all to me, my original, his alterations and W. S.'s contribution, with the rueful comment, "See what you can make out of this lot." Some of my time was taken up in writing reports and descriptions of operations and projects. I used to be quite a fluent writer until he gave me a copy of Arthur Quiller-Couch's *The Art of Writing* to read and digest. I read it and re-read it as instructed and it completely stultified my writing for years afterwards. I believe that all subsequent assistants that Mr Clark had suffered the same experience.

The largest proportion of my time was spent in the west, seven months in the first year, very largely on work associated with Great Western Consolidated. I learned a valuable lesson when returning from one of these trips. The interstate flights Perth to Melbourne were made by DC3 aircraft which, because of its limited range, called at Kalgoorlie, Forrest, Ceduna, Adelaide and Nhill. The schedule provided for the plane to leave Perth at 7.00 a.m. and to arrive at Melbourne at 10.00 p.m. On the leg between Kalgoorlie and Adelaide I was sitting next to a very talkative man who had been one of the vendors of the Mararoa leases to WMC, which subsequently floated a public company, Central Norseman Gold Corporation NL, to develop the leases and in which WMC initially retained eighty-one and a quarter per cent. He appeared to have the grievance that he had been made to sell cheaply, a thought prompted I think by the very good profits that Central Norseman had made from the new orebody that WMC had discovered. He was most outspoken in his criticism of Mr Clark who was the subject of his conversation from the time we left Kalgoorlie until we started to descend into Adelaide when he commented to the effect that he had hogged the conversation and as I appeared to know something about mining what did I do for a living? "I am Lindesay Clark's technical assistant," I replied, rather smugly I suspect. His face was a study as a range of emotions swept across it. We finished the rest of the flight into Adelaide in embarrassed silence. Happily Adelaide was his destination.

The following year, 1951, Mr Clark decided that he would tour some of the North American mines with Gerry Breen, the superintendent of Great Western, to study the latest mining methods being applied in the extraction of similar deposits. It was to be quite a long visit, three or four months, and any time I had outside my routine duties I was to spend planning and estimating the cost of an ore extraction procedure for the Copperhead orebodies of Great Western. This was just fine and I spent all of my time both at work and at home in preparing a report on the project which was awaiting him when he returned. It was unfortunate that while he was away there was a decline in Central Norseman's gold production. Development of the newly discovered deposits at the Mararoa and the Princess Royal had not reached the stage where production could commence and ore was being scavenged from areas where mining had previously been stopped because of declining grade. This was accompanied by a slump in the company's share price.

Mr Clark arrived home a less than fit man. He had stopped off in India on his return journey and had contracted a virus that had put him in hospital and taken a heavy toll on his health. Apparently he was unaware of the decline in Central Norseman's fortunes and was very angry when he heard about it. His immediate reaction was to call for me and ask why I had not informed him of this serious matter while he was away. I thought that my reasons were very valid. Frank Espie, the second most senior executive in WMC, was responsible for all of the Western Australian gold operations and as a close personal friend of Mr Clark's, I had expected that he would have been in frequent communication with him while he was away. Apparently this was not the case and I came in for the full brunt of his displeasure which was further aggravated by the poor conditions of his health. "I think I have been very poorly served," he said. "Send me your report on Copperhead." Fears for my future occupied much of my thoughts over the next few days and were raised to crisis point when Ken Forbes, the chief accountant, came in and said, "Hey, what have you been doing?"

My answer I am sure must have been made in a tremulous voice, "I don't know, why?"

"Well, you got a salary rise and you are the only one on the staff who did." I was called into Mr Clark's office a few days later to receive his commendation on the report. By then his general health had improved and things were back on an even keel.

Life pottered on without anything of any note occurring for the next three or four months and then Gerry Breen, the superintendent of Great Western Consolidated, resigned to take a position with the State Electricity Commission of Victoria. During the month's notice of termination of services there was great conjecture on who would be the next appointee to the position. It was considered a plum position as it would be the largest of WMC operations. Naturally it was the subject of discussion in my own household. On one occasion, in answer to a question from Dorothy as to who I thought it would be, I replied that I didn't know who it would be but rather facetiously added that I thought it should be me. When Dorothy's derisive laughter had subsided I upheld my statements by pointing out that I had taken the major role in preparing the initial capital and operating estimates and that the extraction method that I had designed was being implemented. More laughter.

It came as a decided shock a week or so later when Mr Clark called me into his office to tell me that he had given deep thought concerning the Great Western appointment and had come to the conclusion that the person he wanted in charge was me and how did I feel about that. Would I accept? To say that I was staggered by the offer would be to put it mildly and I immediately revised the opinion I had expressed to Dorothy by doubting my competence to do the job on the grounds that I had no previous managerial experience of any significance. He sounded a bit exasperated when he suggested that he'd be the better judge of my abilities and to make up my mind whether I wanted it or not. Once again I told him that I would need to consult my wife and discuss with her the pros and cons as I was her principal investment.

The crux of the situation seemed to me to be the effect of inflation on the project. War in Korea had already started inflation on a rapid upward trend, the gold price was fixed at US$35 an ounce and the profit margin on the operation was being steadily eroded, conditions which had probably prompted the previous manager to change jobs. I pointed out to Dorothy the very adverse effect a failure would have on my career whether caused by personal inadequacy or through circumstances outside my control. I felt equal to the personal challenge and believed that we should take the risk of the effect of runaway inflation as such an opportunity was unlikely to repeat itself. Dorothy agreed with me that we should accept the offer and the very real risks that went with it.

To say that the announcement of my appointment after it was approved by the board caused some comment, all of it highly critical, would be to understate the reaction. Some even went to the extent of suggesting that Mr Clark had gone off his rocker. It was of course understandable. There were people in the corporation and its subsidiaries that had been in responsible positions since its formation nearly twenty years previously and here was I, an upstart of only three years with the company and for that time not in a managerial position, being hoisted into the prize job. I tried not to let it go to my head. I knew what I had let myself in for and was sobered by contemplation of what lay ahead. I knew that I had a great capacity for hard work and that I had the innate ability to get the best out of people. I remarked to Dorothy, "If I get half way up the ladder and then it collapses and I land in a heap at the bottom, it won't be for lack of trying."

There was only a few days' delay between the announcement of my appointment and my permanent departure for Western Australia. I had been on Melbourne office staff for just about eighteen months, rather different to the ten years that I had been told to expect when I took up that post. I had a pleasant couple of days with Mr Clark when we drove around looking at two of Western Mining's gold dredging ventures in Victoria. He decided it was time to change the formal relationship between us and proposed from then on that I should drop the formal "Mr" Clark and call him Lindesay. The changeover was easily accomplished for he had existed in my thoughts as Lindesay Clark and Lindesay almost from our first meeting. I have always thought it proper that an age or status relationship should be appropriately recognised and it is ironic that I have never been accorded that note of respect. For as long as I can recall I have been addressed as "Brodie" by young and old, rich and poor, male and female and I guess it will continue so.

The day before I left I was called up to receive my "riding instructions" and the blessing of Sir Walter Massey-Greene, the chairman of WMC. Sir Walter was then well into his eighties but remained a shining example to any young man aspiring to succeed in his profession. He went to work in his office on the top floor of Collins House, 360 Collins Street, seven days a week, refreshing himself at midday with a short nap after his lunch, eaten in-house. He came straight to the point after congratulating me on my new appointment. "I am told that you estimated the capital and operating costs on which this project has been based. They had better be correct. We are heading for the very devil of an inflation. If you overrun you will send the company broke for there will be no more money where this lot came from." The interview was short and terse. I got the message. It was not the best boost to my morale that I have ever had but it certainly put me on my toes. His comments were somewhat softened by his final remark, probably an afterthought, that he "felt sure that I was equal to the task".

Dorothy had enjoyed her eighteen months in Melbourne but was pleased to be returning to sunny WA and her many friends. I decided to travel west in one of the Australian coastal vessels which would give me a nice break for a week. There is nothing like sea air to clear the lungs out and as far as I am concerned, nothing like some bad weather at sea to clean out the stomach also.

Chapter 9

I had not had a holiday since I joined the corporation and the time was not so pressing as to disallow me from travelling to Perth by sea. In any event as we were travelling with four children, number four child Charles having been added while we were living in Melbourne, and had considerable luggage and household effects, it was substantially cheaper. As our travel expenses would be to the account of Great Western I was heeding Sir Walter's advice from the outset!

It is an interesting experience to take up for the first time the top position in an established organisation. Although I was not a complete outsider and eighteen months at Central Norseman helped, I think my term at head office was a minus. People in the field never think very highly of head office staff and it was with some caution that I introduced myself to those already on the job, among whom there was one disgruntled chap, the underground manager, who was of the firm opinion that he should have been appointed to the position.

Construction of homes, begun early in 1951, was already well advanced and the five senior staff houses for the superintendent, underground manager, engineer, plant superintendent and accountant had been completed. They had been built on a little ridge adjacent to the main shaft, treatment plant and power house which was very convenient for people filling those positions who might, with the exclusion of the accountant, be expected to be on hand at any hour of the day or night. About a mile away was the site for a new village of three hundred and twenty houses and single quarters and mess where the rest of the team were housed; a mistake which had the seeds of a "them and us" relationship which I took early steps to nip in the bud.

Equally it is a rewarding experience to be in on the early stages of the development of a new community and have the opportunity to guide it and mould its character. I was very firm that there should be no social stratification at Bullfinch and that small though the community would be it would have a variety of social, cultural and sporting activities. As the population increased, so the range of personalities widened and because of the close personal contact I developed and maintained with new arrivals, I became aware of their individual interests and capabilities. A keen photographer would be encouraged to form a camera club, a person with a taste for theatre would be helped to get a group together for play reading and so on. A new community needs someone to assist the motivators to get their act together, to construct tennis courts, sports grounds for football, cricket, athletics, swimming pools etc. on an all in together and do it yourself basis. In those days before people became blasé about religion, churches had to be provided for at least two denominations and, for those who had no need of spiritual solace, places of social relaxation were required.

To get these activities established required money and the most usual and popular method of raising it was by organising dances and the grand occasions of the annual balls. This was a few short years before the new trends in dancing developed and there was still a degree of sedateness and elegance on the dance floor. Usually the evening programme consisted of equal parts of what was called "old time" and jazz. The "old time" favourites were the Pride of Erin, Maxina, Valeta, Waltz Oxford, Canadian Barn Dance, Boston Two Step, Polka, Alberts and, on occasions, the Lancers were very popular. The more vigorous dances stimulated the appetite for both food and drink. Except for the annual balls, it was customary to notify all concerned that the ladies were expected to bring a plate ("Ladies a Plate"). Most ladies had their own speciality, which would range through curried eggs or cold chicken to asparagus rolls, sausage rolls, swiss rolls, lamingtons to cream horns. Of course these were days before the take-away era when women and girls knew how to cook.

I have never danced so frequently before or since as during the Bullfinch period. It was not that I liked dancing for dancing's sake but in a small community it is essential that there be no discrimination in the level of support given to the various groups and it paid off in the high loyalty the people gave me and the company. As an added benefit I got to know all of the wives and enough history of each family to be able to make intelligent enquiries about the various family members when I met the breadwinners on my regular rounds of inspection. There is nothing that enhances a man's or woman's self-respect more than knowing that the boss takes a personal interest in him or her. The result was that the seven years I was there were very happy and completely free of industrial trouble on site.

Heaven knows that I needed all the worker co-operation I could get. Inflation during the construction period increased costs by sixty per cent and left a legacy of disastrously increased operating costs. To get the mine into production called for major modifications to the development and construction plan. The main shaft which was to be sunk to thirteen hundred feet was stopped at nine hundred feet to cut pre-production expenditure but which as a consequence eliminated two levels which were to have been driven and developed for stoping before production commenced. The penalty for this saving in capital expenditure was a reduction in the number of working faces and a serious reduction in the ability to control grade, a situation which prevailed for eighteen months before the lag was overtaken.

Some of the measures adopted to reduce costs gave most enlightening results. Any job that could be let out on a piece work basis was contracted out. This was during a period before specialised mechanical equipment was available on hire and all excavations had to be dug by hand, which made it convenient to pay for it by volume. Bending, placing and tying reinforcing rods, pouring concrete, erecting and welding steel tanks up to one hundred and twenty feet in diameter – everything possible was done on piece work except for small elements of the various jobs which were done on wages to provide a basis for monitoring the differential in the two methods. This firmly established that a man did three times the amount of work when paid by results than when paid wages. A considerable saving was made by these means and the piece workers were happy because they made more money.

The prospectus had stated that mine development and the construction programme would commence in March 1949 and would take two and a half years to complete. Lindesay placed great importance on completing and getting into production on time and I designed the works schedule to ensure it would be completed on time.

The total project, which comprised sinking and equipping with headframe and winders a five compartment main shaft twenty-one feet by seven feet to thirteen hundred feet, driving three production levels at two hundred feet intervals, preparing two open pits for production, constructing a five hundred thousand ton per year treatment plant and a seven megawatt power station, together with the associated ancillary buildings and a village of three hundred and twenty houses for which separate funding was arranged, was scheduled to be completed by 31st

October 1952 at an original estimated cost of £1.5 million. It appears a very small sum judged by today's standards but it should be considered against a basic wage for those days of £12 per week. Most construction materials were in very short supply, particularly steel and cement and I spent quite a lot of time and energy in visiting the merchant supply houses in Perth cajoling or threatening, depending on the response I got from the firm being approached. I would leave home at 5.00 a.m. to arrive in Perth two hundred and fifty miles away at 9.30 a.m., leave again at 6.30 or 7.00 p.m. and get back at midnight or thereabouts.

Alex Stewart & Co, our consulting engineers who had the responsibility to procure the larger items of equipment, appeared to have decided that it was not possible to complete the construction by the scheduled date and had given up trying. The first evidence of this attitude appeared when they claimed that heavy duty large diameter (six inch) roller bearings required for the rod mill and ball mill drive shafts were unobtainable within Australia and, as production could not be commenced without them, we should stop harassing them for greater speed with their design and draughting work.

An old friend, Paul McInerney, who was the WA agent for SKF bearings to whom I referred my problem, recalled having supplied the WA Government Railway with a consignment of roller bearings of the size we were seeking two or three years earlier. As luck would have it the job for which the railways had procured them had not gone ahead and the WAGR was pleased to quit them. They arrived on site a couple of days later. Shortly after this episode our consultants advised that the one and a half inch stainless steel ball bearings of which several hundreds were required for the torque thickeners were also unobtainable. Paul found them in a Brisbane workshop (after a busy afternoon telephoning around Australia).

The consultants really thought they had us cold when they advised that the one-eighth inch steel plate required for the main switchboard was unavailable and the electrical contractors, Oliver Nielsen, would be unable to proceed with its fabrication. Plate of all sizes was in very short supply at this time and a large quantity was required for that job. By some happy circumstance we had it in the mine store. I cannot think why the storeman would have bought it other than as everything was in such short supply when it became available he decided to snap it up. Bernie Smith, our Chief Engineer, acted without delay. That plate was loaded onto a truck, taken into Southern Cross for transfer to the first interstate semi-trailer bound for Melbourne with some spare capacity. The driver did not have long to wait to transfer the plate and it was landed in the electrical contractor's yard the following Monday, three days later. After that our consulting engineers must have decided that we were actually going to achieve our completion target and smartened up their performance.

About two months before the completion date it became necessary to borrow some tradesmen from within the Western Mining Group. The morale of our employees was very high, every one working like beavers, long hours and at any time, day or night. One of the chaps we had borrowed was quite derisive of our efforts and loud in expressing his opinion that we had "Buckley's chance of finishing on time." But against his will he got caught up in the general enthusiasm and about two weeks before the deadline he stopped me one morning when I was doing my rounds to declare with a note of excitement in his voice, "By Christ, boss, we are going to make it." And make it we did – with a bit of fudging.

One of the last items to arrive was the live roll grizzly. It came by rail at 4.00 p.m. on 29th October. The start-up target date was anytime before but not later than midnight 31st October. The rolls were taken off the train and the assembly and installation begun forthwith using "Ciment Fondue", a quick-setting variety which someone forgot to get from the mine store before it closed at 5.00 p.m. Later that night one of the labourers was told to scale the fence to the store yard, proceed to a certain point in the yard where a drum containing the "Ciment" would be found. This was accomplished without delay and the installation completed by 3.00 a.m. on the 30th and the tired gang dismissed to report again at midday. It was with a great degree of chagrin that when the unit was started up for testing it danced around on its foundations because in the dark of the night the unfortunate labourer, searching for "Ciment Fondue", had brought away manganese dioxide in error. The situation was recovered by sheeting over the top of the unit with quarter inch plate and the plant started up at 1.00 p.m. on 31st October without the live roll grizzly. Two men with hastily improvised tools raked the ore across the sheeted-over grizzly top into the crusher and kept it up long enough to run sufficient ore through to test the circuit for any faults. By next day the unit had been securely fastened in place using authentic "Ciment" and the plant operated in the manner for which it was designed.

During the construction period, when it seemed likely that our best endeavours would be thwarted by shortages of essential machinery and stores, I would console myself with the thought that once we got into production, all of our troubles would be in the past and I would be able to sit back and relax. I was very quickly disillusioned and soon came to look back on the development and construction period with similar thoughts to those with which one regards last year's holiday.

Despite the most careful control, capital expenditure had blown out to £2.25 million, an increase of fifty per cent over the estimates. The parent company had dredged this money up from somewhere to ensure that the mine was brought into production and not another cent, or penny as it was in those days, was forthcoming. Regrettably the ore reserve grade fell short by an amount almost equal in value to the estimated profit margin and any shortfall in gold production was immediately reflected in production costs. We had no overdraft facility at the bank and, as the wages had to be paid without fail twice a month, any shortfall had to be taken up by the suppliers. We soon found out those firms that demanded payment on the precise date it fell due and those that could be deferred until we had the cash. It was not a happy way to operate, its only saving grace is that it kept everyone on their toes.

Inflation continued although at a diminished rate and, as the price of gold was fixed at US$35 an ounce, the difficulty of making ends meet slowly and inexorably increased. It was inevitable that at some time we would strike a real trough in gold production and would not be able to pay wages. This occurred during 1954 on an occasion when Lindesay Clark was visiting. In some consternation he asked, "What are you going to do about it?" and I told him I had arranged to meet the Premier in Perth the next day and ask him for a grant sufficient to insulate us against a recurrence of the situation and to fund an accelerated programme of regional development which would bring in higher grade reserves at the other mines which, with Bullfinch, constituted the Great Western ore reserve.

The Premier at that date was Bert Hawke, uncle to Bob Hawke who became Prime Minister of Australia in 1983. Bert Hawke was a Labor man in the old tradition. The security of the working man was among his top priorities. At that time there were five hundred and fifty men on the payroll and a total of about fifteen hundred people in the community who were totally dependent on the mine. He was also a rather ascetic person, some people would describe him as a sour-puss, but he was also a man of action.

He listened in glum silence when I asked for assistance to an amount of £250,000, and then after asking some penetrating questions aimed at reassuring himself that we had a future, realising that we needed some money immediately, virtually on the hour, he called in the Under Treasurer, Reggie Bond, and instructed him to make an interim payment sufficient to cover the

Great Western Consolidated N.L.
Crushed arches and lagging 30ft above 19 Loading Pocket. c 1962.

Great Western Consolidated N.L.
Mechanised open cut operations at Bullfinch. c 1956.

Arvi Parbo 1968.

wages and salaries and to transfer it immediately to the credit of the company's bank account at Southern Cross. I used his telephone to inform the mine accountant that he could safely present the wages cheque. The balance was made available as required to fund exploration. Fortunately we never got so close to disaster again although the economics of gold production continued to worsen. WMC's liquidity improved as Central Norseman in which it owned a little over fifty per cent had succeeded in its exploration at the Princess Royal and commenced paying good dividends which were maintained for many years. This was of great importance to Great Western as WMC was then able to help.

The expectation of a significant increase in the gold price was in the forefront of every gold miner's mind and although some members of the WMC board were in favour of terminating the loans which WMC continued to make to Great Western and let it sink or swim, they were in the minority and WMC continued to fund it as required until the end came in 1962.

Towards the end of my term as general superintendent – I left the mine in October 1958 to take up the position of general superintendent of WMC in Kalgoorlie – I had a visit from W. S. Robinson. "W. S.", as he was familiarly known throughout the mining world, had formed WMC in 1933 and although he was no longer officially involved in the company he visited its subsidiaries and associated companies on the odd occasion. He had been one of the initiators of the concept expressed in the formation of Great Western of basing the company's operations on a number of mines within a region with two or three mines in production simultaneously to give greater security over both tonnage and grade rather than depending on one mine alone, and he had more than a passing interest in us.

I was quite unaware of what an important figure he was, both nationally and internationally, and after one of his visits he wrote a very fine letter to me praising me on my managerial skills, the control I had over the operations, and the vision expressed in the plans I laid for future development. It was a very gratifying letter when received but true to his form he had built in a sting in the tail. "Never," he concluded, "have I seen £3$^{1}/_{2}$ million lost with greater efficiency." It would have been a very good letter to have preserved but without contemplating a possible future use for it I consigned it to the waste paper basket.

In spite of all the problems, the seven and a half years I spent with Great Western were among the happiest and most satisfying years of my life. Not the least among the good things was knowledge that I developed a very happy community socially and industrially. I still meet ex-employees who say nostalgically, "The happiest period of my life was the time I spent at Bullfinch."

During this period a young Estonian named Arvi Parbo applied to WMC for a job and was posted to Great Western as assistant surveyor. It is interesting to note in view of his later outstanding success that when he was asked why he had chosen to seek employment in the gold mining industry rather than the highly prosperous silver-lead mines at Broken Hill, his stated reason was that he considered he would get better management training in an industry that was operating under stress than otherwise. At that time feather-bedding was rife at Broken Hill where, among other things, the union-imposed rule was for two men to be employed on each rock drilling machine and other inefficient work practices were in force while one man per machine was the rule in the gold mining industry in WA.

Arvi very soon made his mark moving rapidly up the staff ladder at Bullfinch and, when some three or four years later Melbourne office asked me to recommend a person for the position of technical assistant to the general manager (W. H. Morgan), he was the obvious choice. The position had not been filled since I occupied it in 1951 and subsequent appointments were intended to be for a twelve month term. When I nominated a replacement and requested the return of Arvi, I was told, "Certainly send the new man over but you are not getting Arvi back!"

Frank Espie had passed the retiring age of sixty-five a couple of years before and had hung on because, as one of the three or four people responsible for Great Western being floated as a public company in January 1950, he felt that he had a continuing responsibility to stay closely associated with it until its problems had been overcome.

Once again there was a high level of conjecture and discussion as to who would get Frank Espie's job. It was the senior operating position in the organisation. At that time WMC was quite a small company with an issued capital of £992,175 operating only three mines, Gold Mines of Kalgoorlie, Central Norseman and Great Western, and a small, highly efficient, exploration unit. At that time the WA mining industry as a whole was small and didn't show any real signs of growth until the 1960s.

I was in the running for the job and felt that I had a reasonable chance and that was how it turned out. One senior person on head office staff, Ken Penrose, the chief metallurgist, left in protest as he considered that he had a stronger claim than I. The rest accepted it very well. Another senior man, Bill Blown, the chief engineer, went so far as to assure me that I could count one hundred per cent on his allegiance which in view of a remark subsequently passed by Mr Espie I found very comforting.

The Bullfinch years were exciting years. They were also very demanding. It would be fair to say that at that time WMC's reputation was in the balance. If we had failed to keep Great Western going it would have had very damaging effects on Western Mining which had initiated the company and was its manager. Great Western Consolidated had some very dissatisfied and outspoken shareholders who waited impatiently for dividends that never came. This was due in part to over-estimation of ore reserve grade and in part to continuing inflation which, although running at a slower rate than that initial disastrous explosive burst which was generated by the Korean war, was still a significant factor. The official gold price fixed at US$35 per ounce was ameliorated by the subsidy per ounce paid under the Gold Mining Industry Assistance Act but the cumulative effect of inflation progressively diminished the actual benefit of the subsidy which ceased to be effective when costs exceeded £A22 an ounce.

Leaving Bullfinch was a sad occasion but also a relief from the unrelenting pressure of trying to make ends meet. I realised that both Frank Espie and Lindesay Clark had a greater responsibility than I in the matter of Bullfinch and that there would be a flow-on effect to me in the new position but that I would be freed from the day to day events which were so exasperating. One such was the occasion when a shift boss on night shift in the open pit decided to move an electric shovel for the operator while he was having crib and not having experience with the shovel got it into motion, could not manoeuvre it or even stop it and drove it over a thirty foot bench onto a heap of broken rock. I had only just acquired it (secondhand) as a means of increasing production of oxidised ore at a time when underground production was falling off. As the shovel took several months to repair it was a serious blow. The situation was worsened by the shift boss having a very large family and usually being reliable, if not very bright, circumstances which decided me not to fire him, an act which would have eased some of the rage and frustration that was burning me up.

Adversity sometimes brings out latent talents or characteristics in people. There were two events which confirm the old adage that "necessity is the mother of invention". Because it had been necessary to curtail pre-production development of the mine until production commenced and a cash flow initiated, a substantial increase in the rate of development advance was essential. One of the most critical development needs was to establish a vertical ore pass adjacent to the shaft between the No. 6 and No. 10 level. The normal procedure in circumstances of

Mr William Baker (Bill) Blown
1908-1991 – Western Mining Corporation's chief engineer (photo c. 1974).

such urgency of simultaneously winzing and raising from the Nos. 6, 8 and 10 levels was beyond the manpower resources of the company as, unless the work was done on a continuous three shift basis, it would not have been completed on time.

I conceived a method based on drilling a large diameter (four inch) hole between the levels (which were spaced at two hundred feet intervals) through which air and water pipes, the hoist rope and signalling line could be passed. A hoist mounted on the main haulage level above would raise a stage to the work face from the level below and provide a base from which the miners could work. After the face of the rise had been bored out and charged with explosives and before firing, the air and water pipes, hoist rope and "knocker line" (for hand signalling between miner and hoist driver) were retracted to the level above. After firing which was initiated electrically, the blasting fumes rapidly dispersed through the service hole and work resumed with minimum delay. The four hundred feet of raising was completed in record time. It was a highly successful procedure and was adopted immediately in Kalgoorlie under the name of "gig-raising". Shortly afterwards it was in use in South Africa and Canada, each country claiming a first in developing the system, a claim which was rightly due to Great Western.

Another innovation successfully developed at Copperhead by Ken Denham, the underground manager, and Bill Blown, WMC's chief engineer, was the adaptation of two Eimco loaders to operate in a heading declined at 46% to the horizontal. Eimco loaders had been adapted to operate on lesser gradients but for successful and efficient operation on gradients of 46% required significant modifications to the equipment. The point had been reached when it was necessary to extend mining operations beyond the depth accessible from the main shaft which at that time bottomed at a vertical depth of nineteen hundred feet, one hundred feet below the lowest operating level. As the orebodies dipped at forty-six degrees and were rapidly becoming further away from the haulage shaft with increased depth it was decided to sink an internal shaft parallel to and equidistant from the two ore strands being mined, in a location which would allow the ore and waste, hoisted during the sink and later from production, to be diverted into the ore and mullock pass systems serviced by the vertical main shaft. The lodes, named the Northern Series and the Southern Series, were a few hundred feet apart and roughly parallel.

The centre line of the new inclined shaft was selected by Don Campbell, Western Mining's chief geologist. At the fourteen hundred level, where a chamber was being cut for the sinking winder, a tight narrow vertical shear, hardly more than a crack, occurred in the middle of the projected position for the shaft and parallel to it. That shear, insignificant though it was, was to prove our undoing. I had recently dismissed the underground manager for dereliction of duty. He was a man of wide practical experience and had he still been in office at the time I believe he would have recognised the problems that could and did develop as the shaft was advanced and would have spoken up strongly in favour of abandoning the selected shaft location for an alternative site. Regrettably neither the man appointed as his replacement nor any other person involved in the shaft sink initially recognised the possibility that the shear could develop in width and intensity at deeper horizons to the extent that it could seriously affect the strength of the ground in which the shaft was being excavated.

By the eighteen hundred foot level the shear had intensified dramatically and the question had to be asked whether it was wise to continue. At this point I asked Lou Checker, a very hands-on miner, who was managing the Nevoria Mine, a part of the Great Western operation fifty miles to south of Bullfinch, for his opinion on our problem. Lou had a long history in mining. I had known him since 1934 when we were both employed as miners at the Triton Mine. He had been very successful passing through the ranks to shift boss, foreman, underground manager and mine superintendent. When Triton closed down after the war, Lou was appointed general superintendent of New Coolgardie NL. When Gold Mines of Kalgoorlie took over New Coolgardie in the 1950s Lou became redundant and I was very pleased to be able to take such an experienced man on to the Great Western staff.

After Lou had made his inspection he joined me in my office to report. Lou was a rather tubby man and found it more convenient to support his trousers with braces rather than a belt. He used these to great effect during a discussion to emphasise a point, hooking his thumbs into his braces, stretching them to the limit of their elasticity and then releasing them at the crucial time. “Get out of it, the ground will beat you”, and the braces snapped smartly back. Technically he was correct. From purely engineering considerations of strength of ground the location should have been abandoned and a new shaft started in strong competent ground. From purely practical and financial considerations of time and money, this was unacceptable. The delay in making a new start would have resulted in the ore reserves above the eighteen hundred foot level being mined out several months before new ore could be developed in depth, nor was there sufficient money to pay for the work involved which was additional to the original programme. There was no other option.

The shaft had to continue on its original designed course. With each daily advance the shearing became wider and more intense and ground support had to be introduced. A steel bending jig was made up by Bernie Smith, Great Western's chief engineer, and forty-five pound steel rails, obtained from the WAGR, were formed into the appropriate shape, stood at five foot centres, and socketed into the nine inch by nine inch jarrah sills. Very soon afterwards the spacing had to be reduced to two foot six inch centres and the roof lagging increased from nine inch by two inch to nine inch by three inch. The sills which supported the shaft haulage rails began to bend upwards from side pressure of the shaft walls and had to be adjusted at regular intervals with an adze to the proper profile. The roof lagging broke in places under the pressure which had to be relieved. The voids were filled with compressible material such as bags of sawdust and bales of hay. Fortunately the shear was wet and this practice did not present a fire risk. Eventually the shaft was punched down to the two thousand four hundred foot level where it was found that the ore positions had diminished in size and grade and were uneconomic to mine. A limited tonnage of ore was produced over six or twelve months during which time the pressure on the shaft walls, roof and floor had to be continually relieved. Everyone associated with that phase of the work was also relieved when the area was abandoned. It is my belief that by now, nearly forty years later, the ground will have completely enveloped the supporting structures which were installed and that the opening made would no longer exist.

Despite these very grave problems no major accident occurred either during sinking or over the short period of ore production. I do not recommend utilising Eimco loaders or any other type of loader designed to work on a horizontal surface to load at an angle of forty-six degrees. The best that can be said is that it is quicker and easier on men than hand loading. The Bullfinch (Great Western Consolidated NL) experience demonstrated to me the inestimable value of having the total support and loyalty of the employees. I am proud of the "esprit'" that developed at Bullfinch which was evident in many ways. Although wages and contract rates were on the low side, and had to be made up by stints of overtime, the feeling amongst the men was good. The operation was completely strike free and productivity per man shift high.

During one year, late in the 1950s, the Red Cross Blood Bank toured the region and in response to my appeal to our employees for strong support Bullfinch turned in the highest per capita return for the region. There was a rather humorous aspect to this. I had organised a team to use their vehicles to bring people into the blood bank centre set up in the local hall where we checked off against the list of salaried staff and wages employees the names of those who had presented themselves. Late in the afternoon I recognised that one of our assayers had not shown up. He had been to Kalgoorlie that day and was returning by air that afternoon. I elected to collect him from the plane and bring him in to donate his 500cc. When I picked him up I thought he looked a bit pale but he fronted up and gave his quota of blood and then promptly fainted. This caused some good-humoured but derisive comments until he disclosed that his trip to Kalgoorlie that day had been to have all of his teeth pulled out in the course of which he had lost rather a lot of blood! We had been rewarding donors with a bottle of beer or a large glass of orange juice according to their drinking habits to make up for the blood they had lost. We decided this chap deserved two bottles.

When the decision to close the mine was reached in 1962, I asked a selected number of people to hang in until the remains of the developed ore had been recovered. Those who stayed were promised employment elsewhere in the group. George Kimpton, one of the last chaps to be advised that his job with Great Western was terminated but that a good job was being held for him at Gold Mines of Kalgoorlie from which he had been recruited, declined the offer saying, "I've had a good job waiting for me at Bunbury for the last six months. I could have gone any time but I was one of the first on the job and I thought I would see it out here to the finish." This was a great reward for me.

Expectations of a big hike in the gold price, which would have saved the operation and had been the main incentive for WMC to keep funding the deficits which were in varying degrees an annual event, were not realised. But the policy of keeping the operation going paid off in another way. It kept together a very good workforce, both salaried staff, contractors, and daily-paid employees, who initially were placed in current operations elsewhere but subsequently became core personnel in new operations. Western Mining Corporation by that time had successfully brought the Darling Range bauxite into production through the participation of the Aluminum Company of America in a new company, Alcoa of Australia and the staff ex-Great Western were valuable in staffing up the new company. Later, some found their way into the iron ore operations which WMC was developing in joint venture with the Hannah Mining Company and Homestake Mining Company.

On being appointed general superintendent of WMC in October 1958, I concluded it was in recognition of my efforts and my technical and managerial skills.

Chapter 10

When I took up my new position it was anticipated that there would be a month or so during which I would pick up the threads of the new job from Frank Espie following which, when I felt secure and familiar with its many ramifications, he would depart for Melbourne to take up the position of executive director of the Broken Hill South company.

To facilitate the exchange of information I occupied the same office as Espie, shared his secretary and worked from the board table at the end of his office. I recall I was drafting a letter to the Federal Treasurer about the diminishing value of the Gold Mining Industry Assistance Act, one of the many such that I sent over nearly twenty years, when in the afternoon of my first day Espie received a call from Melbourne office requesting him to fly over for urgent discussion. He left that evening telling me I would have to find my way around as best I could until he returned. He was away a month and on his return, informed me that he was to pack up his household and personal effects and leave for Melbourne immediately. Three days later he left after calling into the office to wish me luck. So much for the breaking in period.

He had said to me when he officially told me of my appointment to the position, "You will find this job much more difficult than I have," a statement to which I readily agreed, commenting on his many years of experience in the mining industry and his detailed knowledge of the job. "Oh no," he replied, "I wasn't meaning that. You won't have a Brodie-Hall reporting to you. I could always be sure that you would keep me fully informed on all significant events at the mine, good, bad or disastrous as and when they occurred. The others never do and I have to dig it out of them. I suspect you will find the same but with the addition of the new chap at Great Western."

And so it proved. Extracting information became part of a game. I would decide to go to Central Norseman or to Gold Mines of Kalgoorlie and would ring the manager in the evening preceding my visit to ensure he was going to be at the mine. On arriving the following morning, most of the answers to my questions were monosyllabic and non-committal. Every jot of information had to

Frank Espie

be dragged out of them like a dentist drawing an impacted wisdom tooth. I very quickly decided the best way was to invite them to accompany me on visits to the various heads of departments – mining, engineering, geology, treatment, accounting – and get my information direct from them. This suited them very well. This passive resistance diminished after a year or so but more by way of the manager leaving me to visit the departments on my own while he prepared himself for the question period afterwards.

As my successor at Great Western, Ken Denham had been one of my staff. He knew what I expected of him and was only too glad to share with me his modest successes or his many worrying problems. It is interesting to me how many people in senior positions in an organisation conceal for as long as possible bad news, passing on only such items as they believe the boss would like to hear. Such behaviour displays a woeful lack of moral courage and I very quickly made it a practice to inform people down the line of what I expected.

The Western Mining job was partly more enjoyable than the worrying Great Western job and partly not. I have often wondered whether I found the greater excitement and responsibility that usually accompanied each upward step compensated for the loss of direct contact with those at the working face and the influence one could exert over their performance. I suppose like almost everybody else I would have liked to enjoy the best of both worlds.

After a quarter century as a gold explorer and miner WMC had decided to diversify. The Exploration Division was considering how best to get into the bauxite/alumina business and iron ore also looked to be an attractive avenue for diversification.

When Australia urgently needed a source of domestic aluminium, bauxite was imported from Indonesia and refined and smelted at Bell Bay, Tasmania as a quick solution in time to meet wartime needs. Subsequently high grade bauxite was found at Weipa in Queensland and Gove in the Northern Territory. Lindesay Clark proposed looking in the Kimberleys where subsequently bauxite was discovered at the Mitchell Plateau but Don Campbell, chief geologist WMC, recommended that a preliminary search be made of the south west of WA. Campbell had recollections of some preliminary exploration in the Darling Ranges that had been done during the war by the BMR and Consolidated Zinc independently of each other which had discovered bauxite but of a low grade and high silica content. Don recommended that that area be looked at in greater detail before committing to an expensive exploration programme in such an unfriendly environment as the Kimberleys. The previous explorers had backed off the Darling Range bauxite because of the high silica content and the low alumina grade. Soluble silica, aluminium silicates, consume a lot of caustic soda which is the main agent used in extracting aluminium from the bauxitic laterite. It is a costly reagent and the Darling Range bauxites with a silica content of twenty per cent to twenty-five per cent combined with a low alumina content of about thirty per cent to thirty-two per cent, looked to be a hopeless proposition.

The first sampling was a rapid skirmish by motor vehicle of those outcrops and roadside cuttings which were readily accessible from main roads and tracks. The first pass was encouraging, indicating a very wide random spread of bauxitic laterite through the entire region, generally in the twenty-five per cent to thirty-five per cent class. The real win was recognition that the reactive silica content was only a small part of the total silica which was mainly present as non-reactive silica sand. Encouraged by this favourable condition, sampling was stepped up on a broadly spaced but regular scout pattern. The grid was progressively reduced in size and work concentrated on the more favourable areas. Before long it seemed likely that, because of the close proximity of some of the areas to existing infrastructure of housing and population, power and water supplies, rail transportation, harbour and wharf facilities, we had a viable project on our hands.

Alumina refining and smelting to aluminium were way out of the ambit of WMC's experience and Lindesay Clark was advised by WMC's London financial advisors, Lazard Bros, to seek the support of the Aluminum Company of America, the acknowledged world leader in bauxite mining and refining, alumina smelting and aluminium marketing. It was good advice. To identify this very encouraging prospect within the WMC corporate structure a new company, Western Aluminium NL, was formed in which WMC had seventy-eight per cent equity and Broken Hill South and North Broken Hill each had eleven per cent. After its formation an analytical laboratory was established at Kalamunda and an intensive programme of sampling

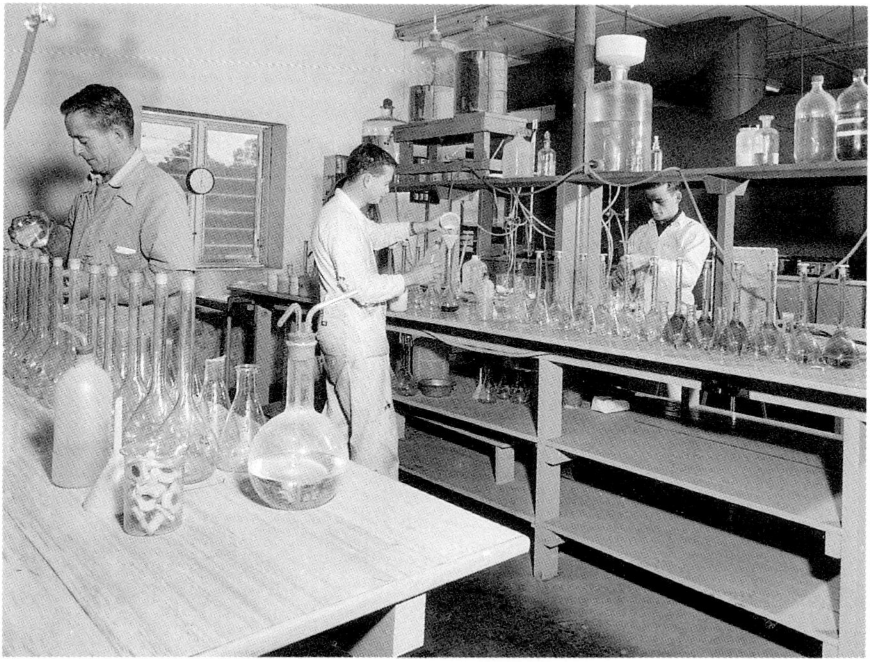

Kalamunda Laboratory

commenced with the objective of proving a sufficient ore reserve to justify establishing an alumina refinery.

Although there was a degree of scepticism within Alcoa of America in the likelihood of such low grade ore being the basis of a profitable business, they agreed to send an ex-employee, Mr Ralph Derr, who had just retired from a top senior position in their alumina processing operations, to come to WA and work in the laboratory to supervise test work and, on the result obtained, make a first pass at determining the feasibility of establishing a profitable bauxite/alumina business in WA. With every passing week Ralph became more sanguine of a successful outcome and his final report resulted in Alcoa committing to becoming fifty-one per cent partners in a new venture. It proved to be a very happy arrangement. They had the money and the expertise. WMC had the exploration skills, personnel and services. Exploration activity and laboratory testing speeded up. Every week saw an improvement in the prospects until, in 1960, the decision was taken to establish mining at Jarrahdale and erect a treatment plant at Kwinana with a capacity of two hundred and ten thousand tons of alumina per year.

During the establishment of the bauxite/alumina project there had been a desire within WMC to get into the iron ore business. I had been a strong supporter of the proposal. The board was far from unanimous of such a move but there was a world-wide shortage of iron ore, production of steel had escalated exponentially during the war, and many traditional sources of iron ore had either been seriously depleted or were nearly exhausted. The post-war demand for steel to rebuild industries throughout Europe, the Soviet and Japan destroyed by the ravages of war was enormous.

Although Australia reported an official reserve of only two hundred and fifty million tons of high grade ore and had embargoed its export, Western Australia attracted the attention of American and Japanese geologists.

Australian geologists had been aware for several years of the great potential for the discovery of large deposits of iron ore in the Pilbara region, which had been observed and commented on by explorers and surveyors in the nineteenth century. Very soon several major deposits were located justifying abandonment of the embargo and setting the scene for a major export business.

The formation by the Western Australian government of its policy for the development, production and marketing of iron ore was heavily influenced by its inability to fund the very high cost of providing the essential infrastructure required. A satisfactory system was evolved which balanced the value of a long term lease

P & H Shovel in open cut. G O J V - Koolanooka.

at a negotiated royalty against the cost to the developing company of providing all of the infrastructure, port development including iron ore stockpiling and ship loading equipment, railway construction and towns with all ancillary services for workers. This called for vast sums of money.

I paid several visits to the area in the early days believing that WMC should take up a position there. Although the vast sums of money required could have been provided by a consortium of foreign companies who would have been glad to get a part of the action in association with WMC, the chairman decided against such a move predicting that industrial conditions would be very onerous and that it would be difficult to establish a workforce of high quality permanent employees in such a remote area with harsh climatic conditions. Perhaps he remembered the recent success achieved by WMC in its discovery of bauxite in the temperate zone close to existing infrastructure and he convinced the board that we should restrict our search to the Yilgarn where some smaller deposits were known to exist. When viewed against the broad picture of the corporation's subsequent history the decision was correct. A major exploration programme in the Pilbara would have overtaxed WMC's Exploration Division which was soon to discover a small iron ore deposit close to the Port of Geraldton and, within a few years, a major nickel deposit at Kambalda practically in its own backyard and once again in a region of developed infrastructure.

Initially the State Government called for proposals from interested companies for the right to mine and market ore from two iron ore deposits, Mount Goldsworthy in the Pilbara and Tallering Peak in the Murchison. WMC made its bid for Tallering Peak and had been accepted when Jim Lalor, a geologist in WMC's Exploration Division, found a better deposit about fifty miles south at Koolanooka Hills. Although it was a small deposit of only seven million tons, a dwarf compared to the Pilbara deposits which were measured in hundreds of millions of tons, it was a short rail haul to the Port of Geraldton and required the construction of about only thirteen miles of railroad to connect to the existing main Perth-Geraldton railway at the town of Morawa where the workforce could live.

The Geraldton Harbour required deepening to accommodate iron ore carriers but, as it was an established port serving an agricultural hinterland, the Government accepted that responsibility. Forty houses were moved by road from Bullfinch and placed on site. The houses, which were timber/asbestos construction, were cut into two halves for transporting and rejoined on site. Two 712 kva English Electric diesel alternator sets were also taken from the Bullfinch power house and re-installed at Morawa. It was necessary to purchase a new diesel electric shovel, a large Caterpillar front end loader and five Haulpak thirty tonne trucks as the Bullfinch equipment was well and truly worn out. Most of the workforce were ex-Great Western employees. These circumstances assisted WMC to earn the honour of being the first company to export iron ore under long term contract.

WMC was not as successful in a deposit it had found at Mount Jackson in the Yilgarn over which I was endeavouring to negotiate terms with the Government. It was a smallish deposit of about fifty million tons at a grade of sixty per cent iron which could have been shipped out of the Port of Fremantle. The company offered to build a standard gauge rail line connecting Mount Jackson with the main Kalgoorlie-Perth line at Northam which would have eliminated the narrow gauge lines running off from the narrow gauge Perth-Northam-Mullewa line which served an important wheat growing district. This proposal, which appealed greatly to Cyril Wayne, the Commissioner of Railways, was unable to get Government approval and, after about a year in trying to overcome political resistance, we withdrew from the area. The WA Government geological survey had published a bulletin on the work of its officers in that area who had failed to discover the deposit. I think their survey of the area must have been made from the comfort of a four-wheel-drive vehicle whereas our chaps walked every metre of the outcrop.

WMC had two American partners in this venture. The Homestake Mining Company and the Hannah Mining Company both had the opportunity to participate through joint venture in the big time game in the Pilbara but didn't take it up. The association with them had some tense moments as they were fully convinced that they had all of the expertise and that WMC were novices, an attitude that diminished over time and was extinguished with the completion of the Koolanooka Hills project on schedule and within budget. When the first train was loaded and despatched from the mine a small ceremony was held at which I had intended to deliver a blast to the Hannah representative, an American vice president named Jack Buford, but I think he may have anticipated my intention because he got in first with a short eulogistic speech praising WMC personnel for all of the imaginary shortcomings of which he had previously been so critical. This did not restrain me from pointing out what a lot of know-it-alls Americans are when they are out of their own country. The passage of words cleared the air and from that point on they left us to carry out our managerial duties without interference other than to visit annually and tell us what a great job we were doing. A personal friendship was established which survived these moments of truth and we have remained close, if geographically distant, friends to this day.

It takes a special kind of mentality or personality to handle joint ventures successfully. In 1960 WMC had the opportunity to joint venture a talc mine at Three Springs. It was owned by a local firm, Universal Milling. They mined the deposit by an underground room and pillar method and processed the ore at a plant at Welshpool. It was high purity talc with especial application in the high voltage electro ceramic industry and for high quality porcelain. They were leaving a very large proportion of the ore behind in pillars and production costs were high. We proposed to extract the ore by open pit methods which, when developed, would permit better selection of ore to meet customers' specifications, give a one hundred per cent recovery of the ore at lower costs.

There were some initial problems on converting from underground to open cut mining, notably from contamination from surface loam and in maintaining a sufficient supply of clean high grade ore during the period when the open pit was being developed. Some of these problems impinged on Universal Milling. It became necessary for them to screen and wash the ore before it could be milled and, until the problems were overcome, board meetings were very unpleasant. Paul McInerney, the man who had been so helpful in resolving our engineering supply problems during the Bullfinch construction period, was the project manager. He had sold his business at the age of sixty-seven to a larger firm and was delighted when I offered him the position of manager of Three Springs Talc. Paul was very much a hands-on action man with a strong element of pugnacity in his make-up which, aggravated by the constant complaints and criticism from our joint venture partners, made my job as chairman unpleasant.

After board meetings each group would retire to different hotels to work their grouches off over a few drinks and I have no doubt that their criticism of us was as bitter as was ours about them. In relation to the size of the company's activities and scale of importance in WMC's business the game was barely worth the candle. Lindesay would often remark that it was not worthwhile WMC becoming involved in small enterprises as they took up as much time of senior engineering and administration staff as major works and returned only a fraction of the profit. But we persisted and inevitably the problems were overcome.

In the open cut, during conversion from pillar mining to open cut. Note the holes in the side of the mine.

Paul made a tour of Europe visiting potential customers for our talc and signed up sufficient business to make it worthwhile to establish a stockpile of lump ore at Rotterdam. This effected a substantial saving in costs over the previous system of selling the product in bags and increased the sales volume many times. With these and later successes relations between the joint venturers improved to the point where these partners also changed their criticisms to songs of praise. WMC subsequently bought the other partner out and, at one time, Three Springs Talc Pty Ltd yielded a higher return on capital and also per person employed than any other business within the corporation.

Chapter 11

One of the most exciting events that occurred during my time with WMC was the Kambalda discovery.

After the incorporation of WMC Ltd in 1933 it remained a relatively small mining house based solely on gold until the late 1950s. Its success in proving the existence of large tonnages of economic bauxite in the Darling Ranges and in bringing the Aluminum Company of America in as major partners in an Australian company, Alcoa of Australia, in which WMC had a significant interest, was its first real step up the ladder to become one of Australia's leading mining houses. The Kambalda discovery coming shortly after consolidated that rise.

At that time WMC exploration operations had only a small staff based at Kalgoorlie and these gains had imbued them with an enthusiasm and dash rarely experienced in a mature company. They could see great opportunities and were willing and eager to take on the world and perform prodigious tasks to push the company ahead. Some quite vigorous discussions had been going on in the Exploration Division concerning where its activities should be concentrated which prompted Roy Woodall, assistant chief geologist, to send me a letter in September 1964 containing the following remarks, "I am not suggesting that I have found an ore body but the occurrence is of great interest. . . . It is fair warning that we should not spend all our effort looking for base metals in the more fashionable areas of Eastern Australia while we have good hunting grounds in our own backyard and still very little competition."

The prospective ore body to which he referred became Kambalda, fifty-five kilometres south of Kalgoorlie. Roy had some of his geologists routinely mapping the district for some time but the discovery was accelerated by information given to him by George Cowcill, a prospector. George had prospected the area ten years before, during the uranium boom, and had found some heavy, mineralised rocks which he hoped might contain uranium, but which Bill Cleverly, lecturer in geology at the WA (Kalgoorlie) School of Mines, found to be iron gossan with traces of nickel. When WMC geologists were active in the area ten years later,

Cowcill, who had teamed up with another hopeful, John Morgan, collected some samples from the same spot and showed them to Woodall. When Roy visited the site a few days later he recognised the significance of the outcrops and posted a small team in the area for three months to conduct a geological reconnaissance of the area. The results were sufficiently encouraging to justify a geophysical survey of an elliptical gossan outcrop. The next phase was under the direction of Antun Triglavcanin, WMC's geophysicist. Antun's results confirmed a strong domal structure and indicated nine geophysically anomalous zones at intervals down its flanks.

Roy wished to drill these immediately. His exploration budget had already been approved. Money was very tight at that time and he sought my assistance to obtain approval for this next stage in testing what had become a very exciting prospect. I put this to the board on 26th September 1964. It wasn't all that easy to prise out of them the $26,000 Roy required to drill three holes but they did agree that if Roy wished to substitute the nickel prospect for any of the items in his previously approved budget it was up to him to decide priorities. Drilling commenced as soon as a team could be put together and on 28th January 1966 the first hole intersected massive sulphides assaying eight point three per cent nickel over a length of nine feet at a depth of four hundred and twenty feet, a tremendous result. When WMC announced its plans to develop the deposit, Morgan and Cowcill each received an ex-gratia payment of $25,000 for their part in the discovery.

Sir Lindesay was adamant that the prospect must be developed and brought into production as quickly as possible, a risky decision on the result of one drill hole, but worth taking on the grounds that following this initial discovery of nickel a high level of exploration would follow and Kambalda would not be the only deposit that would be discovered and WMC should establish its markets well ahead of other potential competitors. The market, at that time, was short of nickel to the extent that spot sales were being made at $10 per pound but there was no saying for how long that condition would last. Taking him at his word, a headframe and winder, air compressors and buildings were brought in from other mines within the group and the main shaft collared and sunk to fourteen metres before results from the next round of drilling confirmed that we had not one ore body but at least two.

I had appointed John Oliver, a thirty year old graduate from the School of Mines, to the position of general manager. He had been my technical assistant for the preceding twelve months and

Miners at crib, Silver Lake shaft.

I was impressed with his ability. He lived up to my expectations. I think there has been no other project that has been developed to production at such speed. Ore production commenced in June 1967, and was processed through a plant with a capacity of one hundred and thirty thousand tonnes per year to produce five thousand tonnes of nickel annually. To complete the work on time it became necessary to fly in a very large mill drive electric motor from England.

Arvi Parbo, who was deputy general superintendent resident in Kalgoorlie, and who, incidentally, has a well developed sense of humour, used to refer to this time as a period of "chaff bag accounting!", implying that there was a chaff bag full of money from which department heads would grab a fistful of money whenever there was need to pay a bill or buy a piece of equipment. Of course it was not like that, but it highlights the pressure that was on everyone to get into production at the earliest possible time. There was no time to put contracts out to tender. There were certain firms that could be relied on to do a first rate job for a fair price and achieve a delivery date. The name of three of these companies remain in my memory, Forwood Downs, Hoskins Foundry and Keogh's, a Kalgoorlie engineering firm.

This great effort resulted in the first consignment of concentrate being shipped out of Esperance in August 1967 on the *SS Botany Bay*, eighteen months after the result of the discovery hole was reported in February 1966. The port was not equipped at that time with suitable loading equipment and an elevating conveyor system had to be improvised for the occasion. Bill Morgan, WMC's general manager, negotiated a long term contract with Sumitomo for the sale of forty thousand tonnes of contained nickel over ten years, the first of many which he steered safely through as production increased.

While mine development and treatment plant construction was going on at a hectic pace, new ore targets around the dome were being drilled at an increasing rate as more drills became available and, by the end of 1967, several new ore shoots had been identified which generated sufficient confidence for an immediate expansion of production to twenty thousand tonnes of nickel yearly, five thousand tonnes to satisfy the Sumitomo contract, and fifteen thousand tonnes to be refined to nickel metal at a refinery yet to be built at Kwinana.

Belief in the potential of the discovery was such that as early as 1966 downstream processing was being considered. In the normal course of events smelting would be the next step to produce a matte for further refining electrolytically. WMC wished

Kambalda Nickel Operations. Aerial view of nickel concentrator.

Sir Lindesay Clark 1896-1986 – Western Mining Corporation

Visit by HRH Princess Margaret.
(L-R): HRH Princess Margaret, Mick Cotter MHR for Kalgoorlie,
LC Brodie-Hall WMC

to position itself in the international market as a supplier of nickel metal as early as possible and decided to short circuit the smelting phase by utilising the Sherritt Gordon hydrometallurgical ammonia leach process to produce fifteen thousand tonnes of nickel powder a year.

The refinery, for which design started in 1967, was completed in 1970 at a cost of $40 million, but by this time the intensive drilling around the dome (twelve drills working round the clock) had shown Kambalda's real potential. Other ore shoots around the dome were either in production or in an advanced state of development, the mill throughput was being enlarged to an annual capacity of one and a half million tonnes and planning for the construction of a smelter with an annual capacity of twenty thousand tonnes of contained nickel was under way. It was designed, built and in operation in 1972.

By then the company was producing forty thousand tonnes of contained nickel annually. It was a marvellous effort and a most exciting development to be associated with. I negotiated with the Government the terms of two agreements, the first covering the construction of the refinery and the second the construction of the smelter. I have described in another chapter the problem I created for myself by conducting the negotiation for the Nickel Refinery Agreement with the Hon Arthur Griffith, Minister for Mines when I should have dealt with the Hon Sir Charles Court, Minister for Industrial Development and Deputy Premier, but I offset this against the fact that in both agreements I succeeded in getting a temporary exploration reserve of five thousand square kilometres. The reserve incorporated in the Nickel Refinery Agreement proved to be particularly valuable. Several important nickel deposits have been discovered within its boundaries by WMC. The other which was granted in the Nickel Smelter Agreement was in the Ora Banda district and proved to be not so useful.

The workforce required to meet this steep growth in production peaked in 1973 when the population reached six thousand. A township had been laid out at Kambalda in 1897 to serve the Red Hill gold mine upon which WMC first built its houses. Due to the hilly terrain which limited expansion to the south, the discovery of new ore bodies immediately adjacent to the gazetted town, and Lake Lefroy to the east, this site quickly proved inadequate and a new site was selected six kilometres away and named Kambalda West. The board had set very high standards for the town. Sir Lindesay in particular insisted that, as far as was practical, it should be in harmony with the environment. Great success was achieved in this.

Visit by HRH Princess Margaret with LC Brodie-Hall (centre) and unknown miner.

Included in a brochure compiled by Gilbert Ralph in 1992, titled "A Pictorial History of Kambalda", are the following remarks,

"From the very outset great care was taken to preserve the natural bushland. The company adopted a policy recommended by its consultant landscape architect, Jean Verschuer, that the town and operations be established within the existing indigenous landscape. This policy set the framework for the development and John Oliver quickly established a fearsome record by dismissing anyone who removed trees unnecessarily."

"He was certainly encouraged by Laurence Brodie-Hall and received the full support of the directors. Jean Verschuer had a special interest in arid lands and in addition to the valuable advice she gave to the company she provided practical guidance to local residents on the ways and means of maintaining native gardens."

As housing lots were surveyed, all trees and shrubs taller than one metre were identified and care was taken to locate houses and other structures so that trees and shrubs were not removed unnecessarily. Kambalda became a model of good practice, so much so that Dr John Beard, a former Director of Kings Park wrote of it in an article in 1970 under the heading of "Kambalda, An experiment of Gracious Living". He wrote,

"Kambalda gives the impression not of a raw mining camp but of a dignified township established much longer than its four years, set in trees and with flourishing gardens. Above all it gives the feeling of a pleasant place in which to live."

Another article in the *National Development Quarterly* of December 1971 titled "Kambalda, Pioneer Australian Nickel Town" stated

"It is possible that in the future Kambalda will be remembered just as much for its pioneering efforts in conservation related to township planning as it will be for its having been Australia's first exporter of nickel."

Kambalda nickel operations have gone from strength to strength but I think that in no other period of its now twenty-seven year life was there ever as much excitement as in those first few years.

Chapter 12

After my appointment to the position of general superintendent WMC in 1958 Lindesay Clark decided that I should have the benefit of a tour of selected North American mines to broaden my knowledge of various mining and processing methods by first hand inspection. He strongly held the opinion that on such a trip it was most desirable if not essential to have a companion. He had made similar trips on his own account and had suffered the lack of company during travel and at the weekends in unfamiliar places where the locals may not have been interested in providing entertainment. Bill Blown, the chief engineer of WMC, had accumulated three months of long service leave and volunteered to spend this in the interests of the corporation by following up on overseas engineering trends, if in return his fares and accommodation were paid. This was agreed to and it proved to be a very satisfactory arrangement. This and later overseas trips indicated that the best arrangement when making such a trip is to go as a man and wife team. Besides the accustomed company a wife is much better insulation against over-generous hospitality which, when the visitors' itinerary is a sequence of one day visits, can be readily handled by the hosts but is exhausting for the visitors. Often the evening entertainments would develop into an international drinking competition. "We've heard how you Australians are one hell of a hard drinking 1ot", was often the concealed challenge casually dropped in the conversation in reply to which one can deny the charge but, in the event, national honour has to be defended. On one occasion when some of the locals were beginning to flag at about 11.00 p.m. they brought in a couple of reserves who had been fishing in a nearby lake.When these occasions were followed, as they usually were, by early rising prior to catching an early morning flight out to the next mine or smelter, one longed for the weekend to come and with it time to recuperate from the wear and tear on the system. These international visits are of great value in keeping up to date with modern trends and establishing a basis for comparing standards of efficiency. I subsequently had frequent overseas trips but, whenever possible, took the precaution of taking my wife with me.

On this first occasion, having been alerted in November 1958 to a pending overseas tour, I had written to my dear old mother, who I had not seen since 1935, telling her I would be in London in May. Her pleasure in this prospect knew no bounds as her weekly letters demonstrated and, although she had the joy of anticipation, she died of a massive heart attack in March a few weeks before I arrived. She had taken great pride in my achievements.

☆ ☆ ☆

Concurrent with the expansionary activities associated with getting into iron ore and bauxite/alumina there was an urgent need to improve the ongoing operations of Gold Mines of Kalgoorlie or as we all knew it, GMK. WMC was a bit late into Kalgoorlie in the 1930 resurgence in gold mining. Most of the prime leases were held either by Lake View and Star, who had been acquiring a lot of the better leases on the Golden Mile since the 1920s, or by the Great Boulder.

The mineralised rocks of the Golden Mile have been intensely folded into a near vertical syncline which effectively divides the field into two, the western and eastern lode systems. The western limb of the syncline is thicker and more intensely sheared, the lode shears are longer and wider and the western system is virtually the heart of the Golden Mile. During 1934 WMC secured some leases on the eastern system which were brought into production by GMK at the modest annual rate of six thousand tons. After the war, from 1946 through the 1950s, several other mines on the eastern system that had been operating in the 1930s were unable to continue and these were progressively acquired by GMK, each acquisition being accompanied by a small increase in its treatment plant capacity. It was here that the WMC policy of developing a highly professional geological team proved its worth.

The aggregated reserves at shut down of the mines so acquired were approximately five hundred thousand tons at a grade of 7.6 grams per ton which, after careful geological mapping and underground drilling by WMC, yielded over five million at a grade of nine grams per ton. But even these additional ore reserves and the increased production rate, which by 1960 had reached five hundred and twenty-five thousand tons per year, were inadequate to withstand the prevailing economic pressures and ensure a long term future and attention was turned to the

Kalgoorlie Mining Associates. Mt Charlotte crushing, screening and loading plant.

Cassidy headframe and surface plant.

Mount Charlotte lease. This lease, which is adjacent to Maritana Hill, was probably part of the ground from which in 1893 Paddy Hannan gathered up the one hundred ounces of gold upon which his reward claim was granted and which focused attention on the Kalgoorlie field.

The Mount Charlotte mineralisation is quite different from the mineralisation of the world famous Golden Mile which has its centre about three kilometres south of Mt Charlotte and occurs in a different horizon of the greenstone series which hosts the Golden Mile lodes.

Mount Charlotte was large, compact and low grade. It was mined intermittently over a period of sixty years from its discovery by fossickers and small companies who concentrated on mining the best of the quartz veins within the ore body for an overall production of three hundred and seventy thousand tons of ore for eighty-two thousand four hundred ounces of gold, an average of 6.9 grams per ton. Later Moonlight Wiluna Gold Mines acquired the ground and did extensive work shaft sinking, winzing and level development in an attempt to prove up a large tonnage of economically viable grade but the results were too erratic for a reliable grade to be calculated.

GMK took over where Moonlight Wiluna left off but with the advantage that technology had improved in the meantime. Diesel equipment was in use in other countries as was automatic hoisting. A new generation of rock drilling machines and tungsten carbide tipped drills had been developed which greatly improved the efficiency of rock drilling. There was no diesel powered load-haul-dump equipment in use underground in Western Australia at that time and whilst on a trip to Canada primarily to inspect the iron ore industry, I detoured to a gold mine at Chibougamou where this equipment was in use. After discussions with the mine management on its operational value I continued down to Tennessee to the factory where the equipment was manufactured to ascertain price, delivery, availability of parts etc. I considered the equipment would be applicable to the extraction of the ore body by the cut and fill method if permission could be gained to use diesel powered equipment underground. An earlier attempt to gain this permission had been made some years previously but had been thwarted by Arthur Moir, a prominent figure in the Australian Workers Union which represented the industrial interests of most of the labour employed in the mining industry in Western Australia. As Arthur had advanced his position in the meantime and was the current Minister of Mines I anticipated some opposition.

I was fortunate to have at that time John Oliver as my technical assistant as he was the son of Charlie Oliver who had also been General Secretary of the AWU in Kalgoorlie and, following the custom of the Labor Party, had offered himself for election to the Kalgoorlie-Boulder seat in the Western Australian Parliament, which he won. I requested and was given authority by both the WA Department of Mines and the Chamber of Mines of WA to negotiate on their behalf with the local branch of the AWU a set of regulations for the use of diesel equipment underground. The discussions arranged through John Oliver, which went on for about three months, were held once a week with the AWU Management Committee in the AWU office at Boulder.

This protocol was quite at odds with what the unions had been used to. During my term as councillor on the Executive Committee of the Chamber there were several occasions when important issues were discussed with the AWU Management Committee. They would be summoned to the Chamber offices after they had come off work, usually kept waiting and when admitted to the august presence of the Executive Council dealt with in a rather summary fashion. I used to feel sorry for these chaps clad in pretty rough clothes sitting in a meeting confronting well dressed executives who, by their position in industry and the local social community, had a big psychological advantage over them. Certainly the industry was under great economic pressure at the time and the Chamber had good cause to contest every claim but it was the manner in which the discussions were held of which I disapproved.

There is generally a humorous side to most things. On an occasion at which regrettably I was not present and cannot personally vouch for this story, the Executive Council were considering a log of claims that had been submitted by the Carpenters Union. Having failed to reach a decision at its regular meeting it reconvened the following morning. After a couple of hours, opinion was still divided when one councillor, Ian Cameron, a dour chap who seldom spoke, taking advantage of a pause in the discussion, voiced the statement, "Christ was a carpenter".

This bald statement of undeniable fact put the request into its proper perspective. After a stunned silence, during which members I suspect confirmed the veracity of the statement by searching their memories of earlier days at Sunday School, they then voted unanimously in favour of granting the union claim. But I digress. My meetings held under different conditions went off well. We discussed regulations in force in Canada and the USA and adjusted them for WA conditions and showed some very useful visuals of the equipment in use elsewhere in the world. By the end of three months we had a set of regulations acceptable to the Mines Department and the Chamber. When I presented them to the Management Committee of the AWU, the Union Secretary, after scanning them through, said, "We wouldn't know if these are good for us or not, we have only got your word for it. My old man said you are alright and I'll take his word for it. We will settle for these." I had worked for his father in the gold room of the Kalgoorlie Ore Treatment Company several years before and had evidently made a favourable impression on him.

Anyway this acceptance was signed there and then and the business was consummated over the road from the AWU offices at the Albion Hotel in a couple of pots of beer. It was an important breakthrough. I did not fare as well with my efforts to reduce costs by the introduction of automatic hoisting. This was well developed in the Swedish iron ore industry in which man cages operated similarly to personnel elevators in high rise buildings and ore was hoisted by an unmanned system which automatically shut down either at a predetermined time or when the ore passes were empty. The secretary of the Federated Engine Drivers and Firemen's Union had no objection to an automated system being used provided that it was fully manned by people sitting on their bottoms and doing nothing. Needless to say we did not proceed along those lines.

The introduction of diesel equipment made the recovery of the Mount Charlotte ore economic. Two Sanford Day Transloaders were purchased from the US and a cut and fill system of mining adopted it being expected that, by this method, zones of low grade ore could be readily identified during mining and left behind in the stope thereby enhancing the ore grade and reducing costs. An ore crusher was installed in the ore body below the five hundred feet level which was connected by conveyor to the ore pass and hoisting system. When I informed the mine superintendent of the decision to proceed his immediate reaction was to declare it to be "sheer lunacy" and the project doomed to failure because of the low grade of the ore body. When it was put into practice events proved him wrong and Mount Charlotte survived the economic conditions which forced the closure of all other mines in Kalgoorlie.

The cut and fill method adopted for the ore between the surface and the five hundred feet level was discontinued below that horizon. Open stoping was then adopted, the open stope being stabilised after the crown pillar had been blasted by introducing dry sand fill from the surface. Open stoping was an improvement on cut and fill but suffered from the defect that, after each successive crown pillar was blasted, the sand fill rapidly percolated through the broken ore until a point was reached when the dilution of the ore by sand fill was so extensive that recovery of the broken ore had to be stopped resulting in a significant amount of ore not being recovered. This caused great heartburn at the time but later, at a most critical period in Kalgoorlie's history, proved to be a real bonus.

☆ ☆ ☆

Alcoa's incursion into the Darling Range for bauxite caused a great furor with the well-intentioned but incompletely informed environmentally motivated members of the public at large, who visualised the entire Darling Range being stripped of its forest cover, and the total destruction of Western Australia's most valuable hardwood, jarrah. I then found myself as Alcoa's main spokesman on this very important matter during which I was once described as the great anti-conservationist! The relevant facts that had to be brought home to these well-meaning people was that the economic ore occurred in pods as small as four hectares and seldom larger than twenty-five hectares. To produce at the rate of two hundred and ten thousand tonnes of alumina per year, which at that time was considered to be the extent of the business, only approximately twenty hectares would be mined

Alcoa Board c. 1979 Standing l-r: Sir Archibald Glenn, R.L. Baillieu, Sir Laurence Brodie-Hall, H. McE Scambler, Hugh M Morgan Seated l-r: Sir Wilfred Brookes, George T. Haymaker (M.D.), Sir Arvi Parbo (Chairman), Joe C Bates, J. Darling.

annually. No one at that time could have visualised that Alcoa would grow to become the world's largest exporter of alumina, mining four hundred hectares annually. But the progressive increase in production over thirty years was accompanied by a regular improvement in revegetation.

It was predicted at the outset by the WA Forest Department and the Institute of Foresters that it would not be possible to revegetate the mined areas because of the dense clay that underlay the bauxite which the tree roots would not be able to penetrate to the extent necessary to support the young growth and which would fail from wind throw.

The critics were very nearly right because of the dereliction of the WA Forestry Department who, instead of applying all of the money paid to it by Alcoa as a reafforestation levy, used a large part of it to establish soft wood plantations in other areas of the state. When Alcoa became aware of this the company renegotiated the terms of the agreement exempting it from paying the reafforestation levy and leaving it solely and totally responsible for reafforestation.

Besides the regular revegetation programme the company initiated a research programme into the cause of jarrah dieback and now, after many years, it has developed procedures which hold high promise that, on the land that has been mined and revegetated since the programme, a healthy forest of mixed species including the proportion of jarrah which occurs in the natural forest, will be established. Research indicates that the fungus thrives in poorly drained, constantly moist areas. By deep-ripping the clays to a depth of two metres, achievable these days with the very large mechanical equipment available, and then fertilising and reseeding by hand to avoid all further use of heavy equipment which would re-compact the clay, waterlogging is avoided and with it forest dieback. For its continuing efforts to reestablish the indigenous forest species and in acknowledgement of the high level of success it achieved, Alcoa was honoured in 1990 through listing on the Global 500 Roll of Honour by the United Nations Environment Programme, UNEP. The UNEP citation reads,

"For 20 years the company has carried out a mine site reafforestation programme in the South West of Western Australia. This covers four bauxite mining areas in a unique forest and three associated alumina refineries. This successful plan integrated overall environmental protection with industrial development."

Alcoa is the only company in the world to have been recognised in this manner for rehabilitation of mining areas.

To fund its share of establishing Alcoa, WMC had a share issue in November 1961 in which the staff were invited to participate on very attractive terms. As this had a significant effect on my personal financial condition it is worthwhile digressing here to describe it.

Lindesay Clark had been well aware that for many past years, during which WMC's sole activity had been gold mining, the profitability of which had been squeezed between a fixed gold price and regularly rising inflation, the salaries of WMC staff had been at the bottom of the salary scale within the mining industry. He resolved that he would compensate for this by offering staff participation in a Staff Share Issue and the opportunity to realise a capital gain anticipated to be generated by the Alcoa business. It was an ingenious and generous scheme and every Western Mining salaried staff employee had reason to be deeply grateful for it.

The basis was the offer of one fully paid ten shilling share at par for each one pound of annual salary. The shares had been prepaid by the company which was to be reimbursed over a five year period, by retention of the annual dividend of one shilling per share to which the corporation was committed, and the deduction of one penny per share per month from salary. At the time of issue the market value of the shares was eleven shillings but began a steady rise and after some share splitting reached $95 per share equivalent in 1969 at the height of the nickel boom.

There were successive public issues and accompanying staff share issues but at lesser ratios. I recall the second offer was one share for £2 of annual salary. They were ordinary shares and carried rights to participate in subsequent share issues. In those days I travelled by car to Perth very frequently, it was a long and boring trip of about six hours duration and to pass the time on one occasion when the shares were at their peak I calculated that if a middle manager on £2000 per year at the time the original offer was made and whose salary was annually adjusted in terms of the increase in the CPI had taken up all new issues and all of the rights entitlements attached to shares he had bought, he would have held shares with a market value of $500,000 when the market reached its peak.

Not many people took up all of their entitlement. Some paid them out and took an early profit, others held a proportion and sold some as the price rose perhaps to enable them to buy a motor car or put a downpayment on a house in Perth or Mandurah or Rockingham. Most goldfielders have such an ambition. For my part it put me in a very comfortable position which I have further developed.

Chapter 13

When I joined the 6th Field Company in 1942 it was located at Strawberry, a point on the road connecting Mingenew with Dongara, and was commanded by a Major Kennedy from Ballarat. An accountant by profession, working for a flour milling company at Ballarat, he had given years of dedicated service to the Field Company when it was a militia unit but I doubted his ability to cope with the exigencies of a combat situation and was delighted to find that the second in command was John Parker, ex the WA Department of Public Works, a first class engineer who shortly after the war was appointed Director of the department and ultimately awarded a knighthood for his contribution to the development of Western Australia. Regrettably our paths separated quite soon as John was transferred to another company. I caught up with him again in Balikpapan and cemented our friendship there. After we both returned to WA I saw quite a lot of him. He was quite a few years older than I and on many occasions I benefited greatly from his advice which was backed up by long years of experience in the public service with which I found myself more and more involved. He had a nice sense of humour and a disarming way of letting you know when there was "nothing doing".

One such occasion was after a deputation from the people of Bullfinch had told me of their need of a community swimming pool, the extent to which they were prepared to help in cash and in kind, and had asked if I would talk to the Government about the availability of water for a pool. I fronted up to John Parker to plead their case and waxed quite eloquent in describing the hardships of country living compared with the luxuries enjoyed by their metropolitan counterparts who had the entire Indian Ocean to disport themselves in. The isolation, the dust and the heat were all played up against the background that the people who asked for this amenity were part of the very small proportion of the population who were producers of wealth, earners of export income, while the rest, the city dwellers, produced nothing but existed by taking in each other's washing.

After taking this in he appeared to abruptly change the subject when he told me of a visit he had received a few days earlier from a young man who had just invented a detergent. John asked him if he had a name for his new product and was told that it was proposed to call it "Stuffed". This prompted him to comment that it was a strange name. "Well," said the young chap, "I've got a jingle that goes with it also. It goes like this,

Omo washes clothes brighter
Persil makes your clothes whiter
If you can't get either
Get 'Stuffed'."

It was not the answer I had hoped to get but I knew then what my chances were of getting the water I needed, at least in the short term. We did in fact get the water. But I had to play politics and invoke the assistance of Lionel Kelly, the local Member of Parliament, who was also Minister for Mines. By the start of the next summer Bullfinch had a swimming pool of Olympic length but somewhat lesser width. Being thwarted in his efforts to conserve the state's water supply for other things which he probably considered more important didn't upset our friendly

L to R: Myself, Lionel Kelly Minister for Mines, unknown, Frank Espie at the opening of the Bullfinch swimming pool.

relationship which was just as well for me as, about fifteen years later, I found myself in a very sticky situation from which John rescued me.

I had unwittingly painted myself into a corner when negotiating with the WA Government the conditions under which WMC could build a nickel refinery at Kwinana. I had previously satisfactorily concluded an agreement with the Government covering the export of iron ore from a small deposit at Koolanooka Hills for which the terms were negotiated between me and Arthur Griffith, the then Minister for Mines. These had proceeded very smoothly and WMC was well satisfied with the terms of the agreement. When Arthur Griffith asked me if I would like to handle the Kwinana Refinery Agreement with him or with Charles Court, Minister for Industrial Development (and Deputy Premier), I believed from his question that I had a choice in the matter and replied that I was quite happy to deal with him. Negotiations of this sort are inevitably very protracted, being delayed by the necessity of a succession of drafts passing between the company's solicitors and the Crown Solicitor who vary the conditions to enhance or protect the interests of their principals. Although we had started discussions in August the agreement was not concluded to the satisfaction of both parties until early December despite the Minister knowing it was essential for the Bill to be enacted before the Parliament went into recess so that the ground selected for the refinery site could be acquired in time for construction to commence in March.

The final draft agreement was presented by Arthur Griffith to Cabinet at the penultimate meeting in December with the request that it be approved as a matter of urgency and passed through both Houses before Parliament rose about ten days later. From this point on I can only recount what I was told occurred at that meeting. Charles Court, who professed to know nothing about the agreement which I very much doubt as Charles had his ear very close to the ground and knew practically everything that was happening or was planned to happen, glanced at the agreement and then said, "What's this, Arthur?" On being told that it was an agreement between the Government and WMC to build a nickel refinery, a project which by rights was in Charles' domain as Minister for Industrial Development, he said, "You ought to know better than this, Arthur, you can't expect an important document like this to be passed by Cabinet practically sight unseen, with only a cursory discussion of the conditions. This is far too important. This will have to wait for the next session of Parliament." Charles carried far too many guns for Arthur and had no difficulty in having the business set aside.

This was disastrous for me. WMC needed more than one hundred acres of land held under a number of titles by numerous individuals which would best be acquired by the Government using authorised powers of resumption if market prices were too high. Had WMC gone into the open market to buy that land, the asking price would have been astronomical. I was brooding over this early in January, wondering whether it would be possible to buy the land at an acceptable price by commissioning a real estate agent to act in secret on WMC's behalf. I had decided it was not a practical measure as inevitably word revealing the identity of the purchaser would leak out, when I thought I would ring up my old friend John Parker and tell him my problem. "If you have a spare hour or so you could come up here straight away," he said. On arrival after the usual platitudes he said, "Now sit down there and I will tell you the background to your problem."

Then followed a comprehensive description of the allegiances and vendettas between family members of the upper echelons of the Parliamentary Liberal Party from which it became clear that the Courts and the Griffiths were not the closest of buddies. "You made the fatal mistake of making the agreement with Arthur instead of Charlie," he said, "but the situation can be recovered if you can eat a large slice of humble pie. How is your appetite?" I told him if it would get the situation back on the rails again I would eat the whole pie with Charlie Court standing over me in public view in Forrest Place. "Righto," he said, "if you've got another hour." With that he rang Charles Court whom I suspect he had primed up while I was on my way into see him and said, "Charles, I've got Brodie here, he would like to talk with you. Can you see him? . . . straight away? . . . Good! He'll be down there in five minutes."

Charlie was all smiles, his prey delivered right onto his door step. After he had enquired after my health and commented on the weather, came the invitation to confession. I stated my case, the penalty that would accrue to WMC by the delay in its construction programme caused by the failure of the Government to enact the Bill for the Kwinana Nickel Refinery in the closing session of 1967, the urgent need to purchase the land for the construction site, etc. At the conclusion, the storm that I had seen gathering while I was commenting on the Government's tardiness in dealing with the Bill burst on my defenseless head. Didn't I know that he was the Minister for Industrial Development? Why was I so stupid as to make the agreement with that lot of incompetents at the Mines Department? For a good twenty minutes I reaped the harvest of my folly and ate huge quantities of humble pie, "Yes, Charles." "No, Charles." "Of course, Charles."

Official Opening Day.
L-R: John Copping (partly cut off), Hon Sir David Brand, Premier, L. C. Brodie-Hall, WMC Director, Hon Charles Court, Minister Industrial Development.

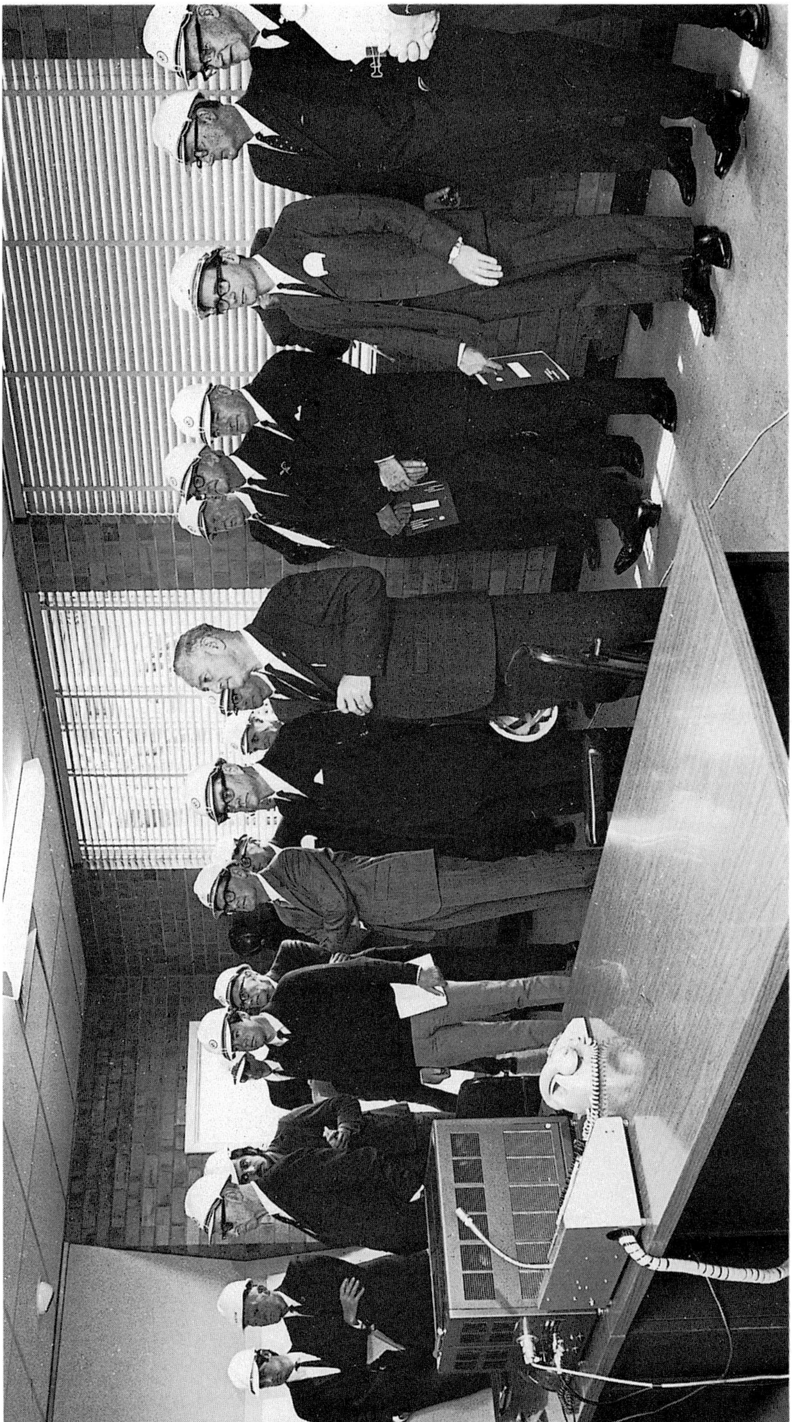

Official Opening Day.
Hon Sir David Brand (centre without helmet).

When he could see that although I could still chew I could not swallow he decided to give me absolution. "Now," he said with a disarming smile, "let me see how I can fix this up." Before I left he had dictated a minute to all relevant Government departments informing them that the responsibility for the Kwinana Nickel Refinery Agreement had passed into his control, where it should have been in the first place, and that pending the introduction of the Bill in the next session of Parliament he was taking immediate action for the acquisition of the land, utilising when necessary the Government's powers of resumption. From that time on all went according to plan. The land was purchased at an acceptable price, the Bill was enacted and construction of the $45 million refinery commenced on time. Importantly, I recovered from the severe bout of indigestion I had suffered from a surfeit of humble pie.

Although we did not see eye to eye on all matters, Charles proved in the majority of cases to be a powerful ally and strong supporter. It was a disappointment to me when he would not support a plan to export iron ore from Mount Jackson in the Yilgarn via the Port of Fremantle. WMC had proved up a deposit of about fifty million tonnes of iron ore and proposed, for the right to mine and export it, to build a standard gauge railway from Mount Jackson to connect at Northam with the newly built standard gauge line from Kalgoorlie to Perth. I saw it as an important step in increasing the standard gauge rail coverage within the state. Supplemented by some road transport and in places third rail construction, it could have serviced the important wheat growing region bounded at the corners by Northam, Goomalling, Dowerin, Koorda, Bencubbin, Mukinbudin and Merredin. Cyril Wayne, who was the Commissioner of Railways at the time, supported the proposal as it would have substituted for very many miles of under-utilised narrow gauge line that was costly to maintain and operate. But Charles had other ideas and our plans were frustrated.

Some years later when Kambalda was being brought into production, I proposed to him the construction by WMC of a standard gauge link between Kalgoorlie, Kambalda and Esperance to facilitate and reduce the cost of transporting nickel concentrates to Esperance for shipment to Japan and of nickel matte to Kwinana for refining to nickel metal. With the services of George Wright of Jackson McDonald I had prepared a draft agreement setting out the terms under which WMC would finance the project. The existing narrow gauge line was routed through Coolgardie and my plan made necessary the construction of a standard gauge line which bypassed Coolgardie, passed through

Kambalda and picked up the existing narrow gauge line at about two miles south of Widgiemooltha from where it would have been widened to Esperance. In return for this and the provision of locomotives and rolling stock at a cost of $9,500,000, WMC would receive an ongoing reduction in rail freight rates on both routes.

It was a good deal and good planning which later encouraged the Railway Department to standardise the line from Kalgoorlie to Leonora. Jim Pascoe, who had succeeded Cyril Wayne, sought my opinion on whether the mineral potential of the region north of Kalgoorlie justified the cost of the conversion of narrow gauge to standard gauge. My confidence in the mineral potential of the northern region of the eastern goldfields has always been high and I told him so. My opinion must have carried some weight as the gauge was subsequently converted.

Charles embraced the Kalgoorlie-Esperance proposal but rejected my draft agreement out of hand. "We will negotiate the terms, thank you very much!" When it was negotiated the terms were very similar to those I originally proposed to him. The freight rates were quite fair, being calculated to compensate WMC for its capital expenditure and give an effective freight charge reduction.

Regrettably the heavy traffic over the Kalgoorlie-Fremantle line, from the heavy iron ore component from Koolyanobbing onwards, caused the line to fail resulting in the axle loading being significantly reduced. To offset the additional cost occasioned by the need to run more trains to carry the same tonnage of freight, the Government reneged on the terms of the agreement and raised the special freight rates that had been specified in the agreement. The company had no redress.

☆ ☆ ☆

Practically all of my adult working life has been spent in the mining industry and the greatest contribution I have been able to make to the community and to the nation seems to have been by keeping sections of the industry going in times of severe depression. A statement like that implies that the gold mining industry is important in its own right and that is a view that is not universally held. There have been occasions when it needed continuing subsidies to sustain it and there have been times when, responding to a period of high gold price, it has made very substantial contributions to Australia's export earnings and balance of payments. Although historically gold has maintained its value over the long term, a factor which should have given the gold mining industry very secure foundations, it has had more than its share of short term problems. A good example of this occurred between 1934 and 1972.

When America devalued its dollar in terms of gold in 1932 by raising the official gold price by sixty per cent from US$22/ounce to US$35/ounce, the Western Australian gold industry experienced a second boom. But in the same way that inflation caused by the 1914-1918 war raised production costs to a level at which many mines could not produce economically and the industry contracted, so were the price advantages of the American dollar devaluation quickly eroded by the 1939-1945 war. By an agreement reached at Bretton Woods after the war, the official gold price was fixed in US dollars and remained fixed until 1972. The economic pressure on the domestic industry had been alleviated by devaluation of the Australian currency against the US dollar in 1949 but the Korean war initiated another burst of inflation and profitability again began to diminish.

To some extent this was relieved during the 1950s by the introduction of improved technology, particularly in drilling, blasting and mechanical loading but these measures were not enough to keep pace with inflation and it was perceived that an ongoing subsidy was needed. Lindesay Clark was the main motivator and architect of the scheme which was eventually legislated by the Federal Government under the title of The Gold Mining Industry Assistance Act 1954. Its passage through Parliament was facilitated by the close friendship that Lindesay had with Sir Robert Menzies, Australia's Prime Minister. It wasn't a very generous scheme in that it paid only seventy-five per cent of the cost of production over a baseline figure of £13/10/0 an ounce and was insufficient to fund exploration required to replace reserves as they were consumed. Moreover it diminished in amount as costs rose until it was extinguished when costs reached £22/10/0 an ounce. Eventually the main concern was with having the terms of the subsidy extended and when I became executive director of WMC in 1968, this became a significant part of my duties which further expanded in scope when I was elected president of the Chamber of Mines of Western Australia (Inc) in 1970. I made many pilgrimages to Canberra to plead the industry's case with either the Prime Minister of the day, the Federal Treasurer or officers of the Treasury Department.

I had some success, the subsidy was increased by $12/ounce early in the 1970s, but the one thing that came through very clearly was the total lack of interest in the gold mining industry by the Federal Parliament which, to a man, they all regarded as a mendicant of no value to the economy and which the nation could well do without.

It always puzzled me that they were unable to recognise that the industry was creating new wealth and that the cost of the subsidy was less than the cumulative interest accrued on an equal amount of money borrowed annually by the Treasury and which ultimately had to be repaid. On top of this financial benefit there was the added benefit to the general economy from the services the industry bought.

In retrospect those visits to Canberra were exasperating. In those days, to avoid using working hours travelling, I flew on the midnight flight, colourfully called the Red Eye Special. On one such occasion, having arranged a meeting with Leslie Bury, Treasurer in the Holt Government, I was astonished to find an invitation awaiting me to join him at lunch. This was unprecedented and I pondered the reason for it without arriving at the correct one. It was his way of breaking the news that Cabinet had discussed my business without giving me the opportunity to present my case to him and had dismissed it before I left the west. He could have notified me of this and saved me both the expense and wear and tear of the trip.

But perhaps the very worst example of the level of contempt exhibited by the Government to industry representatives endeavouring to protect the future economic viability of their industry can be demonstrated by a description of how a delegation from Kalgoorlie fared at a prearranged interview with Prime Minister Gough Whitlam in Canberra. The time was 1974 and the Kalgoorlie Town Council had retained Professor Alex Kerr, a highly reputable Western Australian economist, to prepare a case on the State of the Gold Mining Industry as a discussion paper to support an application for a higher level of subsidy.

The delegation consisted of the Hon Arthur Griffith, MLA, Minister for Mines, the Hon John Tonkin, MLA, Leader of the Opposition in the State Government, Mr Ray Finlayson, Mayor of Kalgoorlie, Professor Kerr and me, President of the Chamber of Mines of WA. Never have I participated in a greater fiasco. Whitlam arrived in the room in Parliament House where we were assembled, grabbed a chair which he sat on back to front and launched into a strong criticism of the gold mining industry with such force that it was not possible for any of the delegation to get a word in for the first twenty minutes. It was obvious that his knowledge of the industry was negligible, even to the extent that he didn't know the difference between Kalgoorlie and Coolgardie. The latter had been virtually a ghost town for many years. Eventually his tirade blew out and our spokesman had a chance to present our case. Before he had finished his presentation an aide came in and whispered something in Whitlam's ear. Whitlam promptly got up, announced that he had another meeting to attend, and departed. Air fares were cheaper in those days but I guess the whole exercise including the economist's fee would have amounted to not less than $6000, say $20,000 in today's terms.

The slow decline in the industry resulted in only five mines remaining in production by 1972. These were Gold Mines of Kalgoorlie, the Lake View and Star, North Kalgurli, Hill 50 and Central Norseman Gold Corporation. The future of gold was so depressed that these companies decided to mine out developed ore, recover broken ore stocks and shut up shop. The workforce could then have been phased out by a gradual attrition resulting in a progressive release of men and consequent enhancement of their chances of finding employment in other industries. Unfortunately when this came to the attention of the relevant unions they immediately demanded an increase of $16 per week for all employees. The minimum wage at the time was $57 a week so it was a hefty increase. A meeting of the respective mine managers was hurriedly called to discuss this at which the two members most seriously affected stated it would be unprofitable to carry on at that added cost and would need to cease operations immediately. However, they said they could maintain production until their residual ore was recovered if the unions abated their demands to $8 a week. This the unions firmly refused to do. "If you can't pay, shut down," was their response.

As a final resort I put the case to John Tonkin, Premier of WA. John was a Labor man cast in the old mould. I told him I thought it would cost the Government $300,000 if it would contribute the $8 per person per week that the industry could not afford. He saw the value of dispersing the labour force over an extended period and agreed to put it to his Cabinet when it met on Monday afternoon. It was then late Friday evening. "Give me a paper showing the number of men involved, the residual tonnages for each mine and the aggregate cost." When I got back to my office a quick phone call to the mines involved enabled me to tabulate their needs and the period over which each company would draw down their requirement and I had the paper in his hands first thing on Monday morning. At six o'clock that night he rang me up to say, "That you, Brodie? You've got your money."

The money was called for over a six month period and, before it expired, the price of gold began a slow move up. It was too late to prevent the shut down of Hill 50 and North Kalgurli but it permitted the other three to hang on, temporarily!

☆ ☆ ☆

To help combat the effect of the severe bout of stagflation which was running out of control during Whitlam's term as Prime

Minister, his government introduced the Regional Employment and Development Scheme, commonly referred to as the RED scheme. Essentially this was aimed at creating new jobs by subsidising the cost of labour in return for industry providing the machinery, plant and equipment component of establishing new businesses or increasing productivity of existing ones. Instead of paying unemployment relief a larger sum would be more productively used as subsidised wages.

At the time, the remaining mines still operating on the Golden Mile which, together with Central Norseman, were the last stronghold of the gold mining industry in WA, were desperately short of skilled miners, many of whom had transferred to the newly developed nickel industry which was in full swing. Not only could the nickel companies pay higher wages but being new mines the working conditions were better and there was a much greater degree of employment security. I saw my duty as chairman of Kalgoorlie Lake View and Central Norseman was to endeavour to turn the RED scheme to their advantage and my position as president of the Chamber of Mines gave me the authority to strengthen any application I made. Conditions were in fact very favourable for the preparation of a strong application. Production from the Kalgoorlie Enterprise Mine, which was a part of the Kalgoorlie Lake View group of mines, had only a short time previously been terminated when costs exceeded revenue.

It had been one of the lower grade mines which had been unable to operate profitably under the rapid increase in costs which rose by sixty per cent during the term of Whitlam's Government. Everything necessary for the immediate reopening of the mine was available on site, headframe, winders, air compressors, change rooms, electricity supply, ventilation equipment. The underground mining equipment, locomotives, loaders, rock drills etc. had been brought to the surface for storage, the pumps were still in place and operating. There was an adequate ore reserve which on a "subsidised labour cost" basis would return a modest profit. A school for miners could be established immediately to train men in development and production mining procedures and practices. At the conclusion of their training the successful trainees would be fed into the Kalgoorlie Lake View operation to increase its level of production. Ore that was broken during training would be toll treated at one of the Kalgoorlie Lake View plants that had spare capacity. Half of any profit arising from the operation would be retained by Kalgoorlie Lake View the other half would be credited to the RED Scheme and reduce by that amount the cost to the government.

I was very taken with the concept and enlisted the assistance of Duncan Bell, who at that time was executive recruitment officer with the Chamber of Mines, to join with me in the preparation of a Project Submission for assistance under the Regional Employment Development Scheme which, when completed, I sent to Mr Clyde Cameron MP, the Minister for Labour I think, but whatever his ministry was the RED Scheme was his responsibility. Giving him a few days to digest its contents I made an appointment and presented myself at his office in Canberra. He greeted me rather cautiously, by that I mean it took him a little while to warm up, but we had a good discussion. To illustrate that it is possible to make a miner out of the most unlikely material I told him of my experience as a trainee miner at the Triton Gold Mines in 1934, pointing out that I had first hand knowledge of training programmes for potential miners. This bit of information direct from the horse's mouth failed to impress him. Visually sizing me up he made the penetrating comment "You don't look like a miner to me!" To which I promptly replied "With due respect, Minister, you don't look like a shearer to me."

The meeting ended amicably but he wasn't committing himself to anything that he might be sorry about later and I went away rather depressed that he had not grabbed the proposal with

Jean and I on the occasion of my being created a Knight Bachelor in 1982.

both hands and run with it. Duncan Bell followed it up later and visited Clyde in his Adelaide office and found that Clyde was in full support of our proposal. He had endorsed the front page of the submission "This is the model we want for the RED Scheme." Until the time of our submission most of the REDS support had been in response to applications from local government and such like. I recall that Jean Vershuer, who was Landscape Architect on the staff of the University of Western Australia, applied for REDS support and had forty additional men working on the University grounds for eighteen months. Most of the REDS projects were short term and Clyde Cameron could see that what we proposed would train men for a new career offering lifetime employment in an important industry.

A bit later Duncan met Dr Wilenski, an important advisor to the Labor government, in Perth when he was en route to Kalgoorlie to look at the mining industry's projected labour needs in relation to the government's immigration policy. I do not know what transpired to prevent the proposal being implemented. Duncan, with whom I discussed the matter later, said that it foundered when it got into the hands of the Canberra bureaucracy and then of course, allowing for the snails pace at which things proceed in official circles, the Whitlam Government was ousted.

There was an interesting sequel to these events. Following the award of Knight Bachelor conferred on me in 1982 by Her Majesty Queen Elizabeth, Clyde Cameron sent me a congratulatory letter saying amongst other things, "I don't go along with that stuff myself but I am sure that in your case it is well deserved. I recall when you came to Canberra to follow up a proposal you had made for part of the mining industry to come under the RED scheme, one of my people said to me, "Look out for this chap he's one of the heavies."" This caused me both amusement and astonishment as never have I considered myself a "heavy".

Chapter 14

During 1972 the board of WMC (managers of Central Norseman and Gold Mines of Kalgoorlie) had decided that operations at both companies should cease and sent out notices to the Stock Exchange to that effect. Mine was the lone voice around the board table recommending that by any and every means we should hang in. Sir James Forrest, at that time a director of WMC, led the opposition and whenever I started to speak in support of hanging in he would good-humouredly exclaim, "Oh God! Here is the silvery tongued orator again." Sir James was a lawyer with a very good business head on his shoulders. At that time he was chairman of Australia Consolidated Industries (ACI), Alcoa of Australia, the National Bank and probably some others – a very nice chap, but no miner! I managed to keep the balls in the air until the tide turned in gold's favour.

The Central Norseman resuscitation (I was going to say resurrection but it wasn't quite dead) is one event upon which I look back with a feeling of achievement and deep personal satisfaction. It confirmed my already firm belief that no matter how hopeless a situation appears, if it can be kept alive by any means, there is always the chance for something to happen which will enhance the possibility of long term survival. I refer to this as being in "survival mode". Once it is terminated the chances of getting it going again are indeed remote. My mother had a saying that concisely expressed that philosophy: "Never say die till you're dead."

Central Norseman had once again run out of ore but on this occasion did not have another ore body in the pipeline. In August 1971 the Directors announced that operations would be suspended when all the economically extractable ore had been treated. Development was discontinued and it was anticipated that operations would cease at the end of 1972.

A young geologist named Jeff Fradd had drilled a pattern of holes down hill from the Princess Royal Reef and had got some encouragement in what appeared to be an alluvial horizon about thirty-five metres below surface. To define the limits and grade of this material required an allocation of money to fund a moderately sized shallow drilling programme. This the Board reluctantly agreed to. Progressively the tonnage built up to an amount at which the grade justified, but only just, commencing a small open pit. The ore had some bad aspects. It contained a high proportion of colloidal clay which, had it been introduced in to the Phoenix treatment plant which was still operating on the quartz ore remnants from the Crown and Princess Royal Reefs, would have seriously reduced its throughput. In addition the gold was very fine and flaky and defied wet gravity concentration, preferring to float on the colloidal clay and was lost to the residue dam. Water, an essential commodity in any form of gold recovery, was only available in limited quantities from the Goldfields Water Supply system and there was no surplus from that source to treat the low grade alluvial in the Phoenix treatment plant. There was however an unlimited quantity of very saline water available from Lake Cowan which under some circumstances could be used. I had a vague recollection that salt water and finely divided clay froths vigorously under agitation so I sent samples of the salt water and the alluvium for flotation tests. The result from the laboratory was most gratifying indicating that, without the addition of the proprietary reagents normally used as frothers and collectors, but relying solely on the salt water and the colloidal clay to perform these functions, eighty per cent of the gold content was recovered in a thin muddy flotation concentrate.

There were some useful pieces of equipment in a disused treatment plant belonging to Norseman Gold Mines, a neighbour company which had gone out of production some years previously, which I could see would be useful in treating this deep lead stuff. I arranged to purchase the equipment. One such piece was a small ball mill which was modified and used to reduce the clay lumps to a slurry. At this stage I enlisted the services of Colin Kleemann, WMC's Consulting Metallurgist, who located a number of old Fagergren flotation cells at GMK. Following the construction of an earth ramp, the erection of a rough ore bin, some belt conveyors, pumps etc, all mounted on a steel structure fabricated out of recycled rolled steel sections, and an extension of the power line from the Princess Royal winder room, we were in business. I have a figure in my mind, for the all up cost of purchase and construction, of $110,000.

The other enthusiastic participant was Graham Wilshaw, a young engineer who was not at all put out by the very pre-used plant and equipment he had to work with. The plant was not a pretty sight but it was built in the minimum of time and importantly it worked successfully giving a seventy-eight per cent

Central Norseman, underground showing loco and winch 1965.

extraction which, considering the simplicity of the operation and the low capital and operating cost, was quite satisfactory. Admittedly it was somewhat different from the normal run of mineral flotation processes. It was utilising salt water from the lake with a dissolved solids content of about thirty per cent by weight and produced a watery concentrate of about six per cent by weight. This assayed about eighty to ninety grams per ton (dry weight) and was transported to the Phoenix plant by tanker from which it was fed at a controlled rate into the cyanide circuit.

This may well seem to be making rather a mouthful of a very simple bit of innovative thinking and improvisation but it was what it led to that is important.

In the course of mining the alluvial, some extensions of the Princess Royal Reef were encountered. They had apparently been too low grade to mine underground and recovering them in the early days by surface excavation would have been out of the question. The later development of very efficient earth moving equipment which could be hired in from contractors for specific tasks changed the economics of surface mining dramatically. Although not a great deal of profit was made out of the first open pit, it encouraged the geologist to recommend testing northwards along the strike by surface air blast drilling. This was not enthusiastically supported by all members of the Board. Sir Wilfred Brookes made a habit of enquiring closely into the operating profits the company was making, if indeed anything at all, out of No. 1 Pit but by this time I had been appointed Chairman of the Board of Central Norseman in addition to my position of Executive Director WA (of WMC) and this gave me much needed leverage.

The Princess Royal Reef had not been mined underground north of the main haulage shaft, there being strong indications that the reef had terminated at that point on a very tight shear. The surface drilling discovered shallow high grade reef for a considerable distance north of the shaft. A second open pit was commenced north of the shaft which disclosed reef typical of the North Royal Reef and this in fact was what it proved to be. It was a god-send, readily accessible and high grade. This of course was a compelling reason why the shear should be investigated more thoroughly underground to the north. The Princess Royal shaft was reopened and further development confirmed that the shear had become very tight and that the reef had pinched out on a tight drag fold but that it developed again further north. A decline haulage way was driven from the bottom of No. 1 open pit in November 1977 to gain access to the ore at the No. 5 Level. This was fortunate as a serious collapse occurred in the reopened

North Royal shaft that took several months to repair. To the north the reef blossomed again into a thick high grade reef characterised by a 12mm seam of finely disseminated gold which traversed the full extent of the reef extensions and raised the grade of the entire stope to eight ounces to the ton.

The effect of this exciting succession of events is illustrated by the record of annual gold recovery. In 1975 gold production slumped to twenty-eight thousand ounces; in 1976 it rose to seventy-five thousand ounces, 1977 to a peak of one hundred and one thousand ounces and thereafter settled down to a steady annual production of around eighty thousand ounces.

During 1977 Central Norseman shares rose to $12.50 a share from a low of forty-six cents in 1974. In 1979 each fifty cent fully paid share was split into five 10 cent fully paid shares which then also rose to a peak of $12 a share.

The steady increase in the gold price, which rose from a high of $70 an ounce in 1972 to $715 an ounce in September 1980, was a very significant factor in the high level of profitability which funded subsequent successful development but a high gold price is not much help if there is no gold to sell. Jeff Fradd's dogged persistence in his search for remnants or new deposits was fully supported by Roy Woodall, WMC's Chief Geologist, who I encouraged and successfully defended at the WMC Board meetings. This work initiated the cycle of events which saved the day and gave Central Norseman a new lease of life. It is not often that a company Chairman has the opportunity to get involved in hands-on activities and I was delighted when it came my way.

Central Norseman has been in continuous operation since its formation in 1935 and is the longest continuously operating gold mine in Australia. If the development of the new area referred to by the company as the Golden Rainbow is successful, the Company appears to have many years still ahead of it.

☆ ☆ ☆

At the same time as the Central Norseman rescue, GMK was also feeling the cold breeze. The payment of dividends which had been a regular annual event since 1938 was passed over in 1967 and several succeeding years, and in 1971 the company posted its first ever operating loss of $1,513,248. Although the price of gold started to rise when America repudiated convertibility of dollars into gold, the inflation that accompanied the period of Whitlam's Government more than offset the rising gold price.

Following its discovery of nickel at Mount Windarra, Poseidon purchased the Lake View and Star leases with the intention, as

I understood the situation, of basing their workforce in Kalgoorlie on a fly-in fly-out arrangement, transporting their ore by rail to Kalgoorlie and utilising the eight hundred thousand tonne/year treatment plant and diesel powerhouse on the Lake View and Star lease to get early production from their nickel mine. Recovery of nickel by flotation is not much different to the flotation process applied to the recovery of gold. Kalgoorlie had all of the infrastructure, housing, water supply, amenities etc. and a resident workforce. Unfortunately for the Poseidon directors they were not able to implement these plans as they were forced into receivership and ultimately their nickel leases were bought by WMC.

There is an interesting aside here. When, in 1968, WMC needed $48 million to build a nickel refinery, Sir Lindesay called in his mining engineers and accountants to seek their advice on whether the shareholders should be asked to fund the cost by taking up shares in a new issue, which at the time would be the largest share issue made by an Australian company, whether it should be borrowed from a financial institution or from a combination of both. At the time the nickel price was at an all time high and demand was very strong. The advice he got after his staff had mulled it over was to borrow the money, as under the prevailing market conditions it could be repaid in four years and there would be less shares to service. Lindesay's reply reflected his many years of experience. He said, "I'm a bit older than you, you know" – very much of an understatement as most of those present must have been thirty years his junior – "And I can remember the Great Depression. I don't know when the next one is coming. Conditions won't always be like they are today and they can turn quite suddenly. Sometime there'll be a slump. The price will fall, nobody will want your nickel but the banks will want their money. They won't stick by you, you know, they'll sell you up. We'll go to our shareholders, they'll have to stick by us." The float was a great success and in due course both the price and demand for nickel dropped very significantly for quite a long time. Poseidon chose to borrow the money with disastrous results.

After Poseidon came out of receivership it seemed sensible to me to propose to them that they should amalgamate their unencumbered Lake View and Star operations with GMK in an endeavour to survive by reducing costs and overheads in a combined operation. As a further incentive GMK staff believed that there were similar opportunities to discover high grade ore in the LVS leases to those that had proved so bountiful in earlier acquisitions of Paringa, Boulder Perseverance, Kalgoorlie Enterprise, New North Boulder and others in the 1950s.

The Chairman of Poseidon, Professor Eric Rudd, had served in the same unit as I in the war. In fact we shared a tent for a considerable time and, both being associated with the mining industry, he having held the Chair in Economic Geology at Adelaide University, we became firm friends. To bring this deal to a head I visited him in Adelaide and negotiated with him the terms of a prospective new company to be named Kalgoorlie Lake View Pty Ltd into which GMK and LVS. would transfer their assets and each have an equal share. WMC, which owned about thirty per cent of GMK, generously offered to contribute $1 million for working capital to earn six per cent equity and the agreement was signed in June 1973.

John Oliver was offered the position of General Manager of the new company. At the time he was managing a joint venture developing a coal deposit at Hail Creek Queensland. John was a man who relished a challenge and he accepted but on the condition that he had the option to return to his position with the Hail Creek Joint Venture if the gold price dropped to a level which imposed restrictions on KLV's future.

The plan which was evolved envisioned rationalising administration, mine services, maintenance, production and treatment, and increasing production from LVS leases from the existing thirty thousand tonnes per four weeks to sixty thousand tonnes, the capacity of the LVS plant at which it had operated before the economic rot set in. Production of necessity had to come from the lower levels of the mine, below one thousand metres as although geological re-interpretation of the upper levels where the occurrence of high grade remnants was forecast had commenced, it had not been possible to establish sufficient high grade ore in the time available.

In retrospect I guess I should have realised that at LVS, an operation that had been running down for fifteen or twenty years, the condition of the plant would have seriously deteriorated through long years of inadequate maintenance. The extent to which this had occurred became apparent when we tried to increase throughput.

The Oroya plant of GMK was in somewhat better condition but, because of differences in some features of the two plants, it was reserved for Mount Charlotte free-milling ore and oxidised ore from open pits. The harder refractory underground ore from the combined Fimiston leases went to the LVS plant which proved unequal to the task and the target of sixty thousand tonnes per period was never reached.

The gold price of approximately A$70 an ounce at the time of the formation of KLN increased steadily to A$145 an ounce in December 1974 but it was accompanied by the excessively high rate of inflation (sixty per cent in twenty months) which occurred during the term of the Whitlam Government. The lift in the gold price was temporary and when, as a result of the US selling off some of its gold stocks by auction at regular intervals to defend its currency, it declined to A$110 an ounce at December 1975, there appeared to be no alternative to closing down the Fimiston mines and running only on Mount Charlotte and open pit ore. That meant we would be back to where we were before we took Poseidon in.

Before the 1972 Federal elections (sensing a Labor victory at the polls) acting in my position as President of the Chamber of Mines of WA I had invited Gough Whitlam, leader of the Federal Opposition, Frank Crean, Shadow Treasurer, Rex Connor, Shadow Minister for Minerals and Energy in the Labor opposition and another Labor MP named Jenkins to visit the major mining centres in WA. Whitlam elected to go separately with Lang Hancock to the Pilbara. I took the others on a more general tour encompassing bauxite/alumina and gold as well as iron ore. I got on quite well with Rex Connor who had a hostile view of senior people in the mining industry who had been, as he put it, "born with a silver spoon in their mouth" and had the advantage of the quality of education that goes with it. He had had to suspend or terminate his tertiary education at an important stage because of the hard times his family endured during the depression. He took kindly to me after he learned in general conversation about my career, especially of my early years in Australia, on which he commented with some surprise, "Hm! You had a rougher time than I did." Despite the general feeling against him held by other executives in the mining industry I quite liked him.

I often judge how well I like people by asking myself how I would like to spend a week alone with them on say, a fishing holiday. I applied this test to Rex and decided I could do that with him and enjoy it. His conversation was interesting, he was a good listener, sincere in his desire to do his utmost for Australia's good, and I am sure would have done his share of camp chores. There was an interesting echo to this relationship when my brother-in-law, Professor Ralph Slatyer, Chief Science Advisor to Bob Hawke and later Paul Keating, being aware that I was writing my memoirs, sent me the following note.

"Towards the end of 1991, I was at a dinner in Melbourne to announce a series of National Science Awards. The Governor-

KNS – Official Opening. L-R: Hon J Tonkin, Hon G Whitlam, Sir Lindesay Clark, L. C. Brodie Hall.

Nickel Smelter for Western Mining Corporation. Near Kalgoorlie.

General, who was the guest of honour, was telling some anecdotes to Hugh Morgan and me about colourful political personalities and was talking about Rex Connor when he said that Connor's initially negative view of the leading figures in the mining industry was changed dramatically on his first visit to Kalgoorlie.

"The Governor-General said, 'This came about when he met a top management executive who had worked his way up from humble beginnings and had roughed it over many years. He had excellent relations with his entire work force and was obviously widely respected and admired.'

"The Governor-General couldn't remember this person's name. Both Hugh and I simultaneously said, 'Was it Brodie-Hall?'

'Yes,' said the Governor-General, 'That was the man.'"

It was about that time that I had my first encounter with Paul Keating who visited the Goldfields in company with Chris Hurford. I was at the Boulder Perseverance mine at the time discussing with Jack Manners, who was then General Manager of KLV, future moves to ensure the company's survival. Perhaps Keating and party had come as a response to a request that I had made to the Federal Government for help to tide us over to better times – I don't know if that was so but it might well have been as I usually had a request of that sort either in preparation, in transit or in their hands. Anyway, it was on that occasion that I was told with great force and personal rancour that the Gold Mining Industry was nothing other than a mendicant, bludging on the taxpayer and the sooner it went bust the better!

Labor won the elections and on many occasions when I was in Canberra during his term of office I would call on Rex Connor. He used me as a sounding board for his proposed legislation such as the National Gas Pipeline, being anxious to know how it was received in the West. There was a reciprocal benefit as I got to know what he had in mind. This brought me under criticism from quite a few people in the Liberal Government in WA as well as in Industry. Charles Court made no bones about accusing me of "going over to the other side". The bald facts are that if you want anything in the political area, you deal with the Government, resorting only to the opposition party when you are not getting anywhere. This was the stance I took with both Connor and Crean in Canberra and John Tonkin in the West.

Rex Connor was still the Minister for Minerals and Energy in 1974 and, as a last resort, I decided to make one final effort to try to get the Federal Government to come to the rescue and made an appointment to see him. It was an unfortunate meeting. Very unusual conditions of persistent fog prevented my plane from landing at Melbourne where I was to transfer to a Canberra flight. The plane back-tracked to Adelaide for the night. The following morning Melbourne was still fog-bound and for some reason the plane detoured to Launceston, landing there before returning to Melbourne. Luckily I got a seat on a Canberra plane where I landed at 1.00 p.m., hours late for my appointment. Rex Connor had been delayed by similar circumstances and when I saw him at 3.00 p.m. he was in a foul mood and gave me short shrift. After listening impatiently to my tale he said, "Fold it up. If you can't make it pay, fold it up."

With my last chance of assistance blasted I returned to Kalgoorlie and on 7th November called a public meeting in the Boulder Town Hall to spread the bad news. It was a good meeting attended by about eight hundred people, many more than the Hall could accommodate. The overflow heard the proceedings from loudspeakers on the footpath. The general secretaries of all the mining-industry-related Unions came up from Perth for the occasion. Gilbert Ralph, my technical assistant, had done an excellent job, preparing for screening graphs depicting gold prices, inflation rates, costs and losses sustained over the previous five years. The facts were irrefutable. The Union representatives asked their questions, got their answers and were amazingly docile. The meeting ended with everyone, with one

Gilbert M. Ralph in Melbourne, August 1986.

Boulder Town Hall.

L to R: Sir Arvi Parbo, Sir Laurence Brodie-Hall, Harry M. Conger, Chairman Homestake Mining Co of America, Keith Parry.

L to R: Hon John Tonkin, Sir Laurence Brodie-Hall, Hon W. Grayden, Jack McDermott.

Kalgoorlie Mining Associates recommencement of operations at Fimiston. L to R: L. C. Brodie-Hall, Hon Sir Charles Court, Harry M. Conger, Chairman of Homestake Mining Co of America, 1981.

exception, satisfied that closure of the Fimiston underground mines was inevitable.

The exception was Graeme Campbell, now Federal MP for Kalgoorlie of several years standing. At the time I think he was an AWU representative. Waving a bundle of GMK annual reports of previous years he held the floor for a quarter of an hour or so making wild unfounded allegations against the company in a vain attempt to refute the irrefutable.

I must admit getting a bit of mischievous satisfaction out of the meeting. WMC's general manager of operations, Keith Parry, and Stan Carter, WMC industrial relations consultant, who both felt that this matter was in their domain and should be handled by them, were sitting in the front row waiting for my play to misfire. I wasn't always in favour of the way they handled similar situations. I left the meeting feeling rather smug.

It proved not to be the final shot I had in the locker. In a further attempt to salvage something and enhance the chance to keep breath in the corpus of the operation I tried to get $3 million from Government sources to finance a development program at the Fimiston leases. The Federal Government had just changed from Labor to Liberal with Malcolm Fraser the new Prime Minister.

(Sir) Charles Court, the State Premier, undertook to put in $500,000 if the Federal Government would find the other $2,500,000. It was apparently too soon after the election for the new Government to make up its mind but in a very generous action Charles said he would put up the State's half-million ahead of the Federal decision. This allowed the mine manager to select a good team of development miners before the general Fimiston workforce was paid off. It was mid February before the Federal Government notified me that it would not assist. The State Government then terminated the development programme on which it had spent $440,000, an amount which was repaid to the Government by the company when it became financial. The progressive reduction in the scale of operations, culminating in this final closure, resulted in the workforce shrinking from fifteen hundred a few years before to seven hundred, and this was a severe blow to the Kalgoorlie community.

Neither of the major shareholders was able to put in more capital and, although WMC made funds available to keep the company afloat, it was not prepared to put in more funds by way of purchasing more equity as it was sensed that these were circumstances in which the corporation would be seen not only as Big Brother but as taking advantage of the other smaller companies misfortune. I was forced to seek a new partner.

There had been an ongoing relationship with the Homestake Mining Company of the USA since 1934 when WMC brought out a team of Harvard trained geologists to re-interpret the geology of the Golden Mile. Some of them became senior executives in Homestake, notably John Gustafson and Paul Henshaw with whom I had close dealings during the Koolanooka Hills iron ore project. The Homestake people were gold buffs from way back – the company had been founded on a gold mine at Lead, South Dakota in 1873 – and they seemed to be people who should be approached. Their officers didn't have much interest in the Fimiston leases where the lodes, although numerous, were narrow, short and labour intensive, but were interested in Mount Charlotte which was large, highly mechanised and better able to withstand the effect of the continued increases in labour cost which could be expected.

At this time KLV was very much a hand to mouth operation. The company had $3 million of debt and no cash other than that generated on a day to day basis, As long as production could be maintained the debt could be kept rolling over. If it ceased we would be in dire trouble. As we paid the bills thirty days in arrears, the current periods gold paid debts incurred two periods previously. I was very anxious to find a partner with some money. Negotiating terms favourable to KLV under which the new partner would buy in was very difficult as the gold price had been plummeting and the other party knew our plight. I had no intention of exposing myself to the accusation that I had sold a pig in a poke and thoroughly explained the situation the company was in. The deal was eventually agreed in December 1975. Homestake acquired a forty-eight per cent interest in a new venture to be called Kalgoorlie Mining Associates (KMA) for $3 million irrevocably committed and a further $5 million contingent upon the new association being unable to fund essential development out of cash flow. They could withdraw at any time after the payment of $3 million and would retain a ten per cent interest. The agreement was signed and Kalgoorlie Gold Mining Associates was formed on 10th March 1976. The gold price was then $107 per ounce.

Shakespeare wrote, *"When disaster comes, it comes not as single spies but in battalions."*

How right he was. As a further tightening of the screw on gold, the US Government, to strengthen its dollar internationally, influenced the International Monetary Fund to reduce its gold holdings by one third by returning one-sixth (twenty-five million ounces) to the depositors and to auction off another twenty-five million ounces by regular auctions over a four year period.

The nervousness induced by the previous US Government announcement of a regular auction of gold by the US Treasury was mild compared to the fright that the IMF auction programme caused. By late August the price had fallen to A$83 an ounce and the mine was losing money at an alarming rate. I went to San Francisco to consult with Homestake, stopping en route to ask Prime Minister Malcolm Fraser for some aid to tide us over to better times. The answer was a lemon.

The Partnership decision was to mine out all recoverable ore and then suspend operations. The probable date for termination was the end of period six, 10th December, 1976. Surprisingly the gold price began to rise in the face of massive US/IMF sales with demand generally for three to four times the amount offered, and by mid November the gold price had risen to A$105 per ounce. On 29th November the Australian dollar was devalued against the US dollar by seventeen and a half per cent, increasing the gold price to A$130 per ounce. On 9th December the Management Committee met at the mine to consider a KLV proposal to continue production at a reduced rate and to recommence development. Homestake said it was not prepared to proceed with a partner (KLV) that was not financially sound. It appeared that Homestake was attempting to force acceptance of an offer they had made to Poseidon for its share of KLV. There was some argument, Homestake seemed obdurate and a press statement was prepared at 3.30 p.m. confirming cessation of production. By 4.10 p.m. and before the press statement was released, the Homestake representative had second thoughts and agreed to accept the KLV proposal. Retrenchment notices were withdrawn immediately but many employees had made forward employment commitments and by early January 1977 the workforce was down to ninety-seven and many top class employees were lost to the industry.

Then began a most exciting time. Quite a decent surplus had been made from the recovery of the ore stocks but there was no developed ore remaining. There was however a significant quantity of broken ore which had been left in the stopes when it became too heavily diluted with sand fill, The gold price was rising and showed good signs (to the optimistic) that the rise would continue. A rapid calculation indicated that it would be feasible to draw the diluted ore and screen out the sand fraction. A screening plant was hurriedly designed, fabricated and erected adjacent to the main shaft ore bin to which the lump ore, after screening, was directed. A conveyor was built to take the sand fraction from the screen to a nearby winze connecting to an empty stope on the

Reward ore body and we were back in business. As Wellington said at Waterloo, "It was a near run thing."

Eventually it was necessary to supplement this source of cheap ore by beating back the brows of the mill holes to the stopes as we finished them. Progressively we got back to a normal production routine based solely on Mount Charlotte. From that point on things were on the up apart from the difficulty of dealing with partners on management committees. I was Chairman of KLV from its inception, had acceded to the Chairmanship of GMK from Sir Lindesay in 1974 and was Chairman of KMA from its inception also. Homestake was the biggest problem. Paul Henshaw who had been Chairman of Homestake through the negotiating period became sick and retired. Henry Colen, with whom I had negotiated the deal, had been made the sacrificial lamb and dispensed with when it seemed likely that Homestake, having spent the A$3 million to which they were irrevocably committed, would have to contribute more or water down. The new man Homestake appointed to take his place was totally uninspiring and faithfully pushed his company's line as Caution, Wait and See.

Although the gold price continued to rise during the period, the cash flow of the project was affected by the distribution of gold in the orebody which also made Homestake nervous. Because of the location of the haulage shaft and the mining method adopted, the orebody had to be entered from the south end which was about one gram per tonne lower grade than the mine average of 4.5 grams per tonne. When development of the Fifteen Level was complete there was a three hundred foot deep ore block available for extraction. The early grades averaged about 3.5 grams per tonne and Homestake got very nervous. I think they had no intention of putting more money in and were reaching the point of withdrawing and maintaining their ten per cent interest when we got into the high grade centre stope – 5.5 grams per tonne. What their reaction would have been if they had dropped back to ten per cent before we reached the higher grade ore is food for thought. I am sure they would have believed we had engineered it. From that point on the grade more than compensated for the low grade south stope and grumbles turned to smiles. Well, perhaps that is an overstatement!

In the 1975/76 financial year when Kalgoorlie Mining Associates took over from Kalgoorlie Lake View, an operating loss of $5,462,000 was sustained but was converted into an operating profit of $410,000 in 1976/77. The operating profits for successive years to 1982-83 were $9,985,000, $29,130,000, $27,726,000, $12,448,000 and $43,800,000.

After the Mount Charlotte operation had been restored and the consortium was well on its way to sustained recovery, early stage development of the eighteen hundred foot level encountered a strong east dipping forty-five degree fault. This was a very severe set-back which neutralised the value of the great effort that had been made to get ahead of production and had established a three hundred foot stoping block.

In the normal course of events, without the complication of the fault which for some unstated reason was named the "Flanagan Fault" although I am sure that honest prospector would not have been at all pleased at being landed with the role of god-father to such a disaster, work would have gone ahead deepening the main shaft and preparing a level at two thousand one hundred feet.

Some underground diamond drilling indicated a movement of about four hundred and fifty feet to the east and four hundred and fifty feet down, but as it was not known to what depth the down faulted section would persist or at what size and grade, a deep hole to be drilled from the surface to intersect the displaced ore body at three thousand five hundred feet to four thousand feet was planned. As the hole would commence in what was expected to be unmineralised ground, to save money, the first eight hundred feet was to be drilled by a percussion drill after which it would be continued to designed depth by diamond drill. Percussion drilling through the near surface oxidised ground was fairly rapid and, although the ground was expected to be unmineralised, regular sludge samples were taken. The hole had passed five hundred feet and was drilling ahead when Jim Cleghorn, KMA's Chief Geologist, rang me to tell me that some unexpected values had been encountered and he had sent the sludge sample down for me to have a look at and consider. I should get a panning dish and pan it off. This I did and found the sample to be very rich in coarse gold. There was a fair amount of fine gold present also. Here was a quandary. The intersection was made in real tiger country, a position in the green-stone series remote from both horizons that hosted both the Golden Mile and the Mount Charlotte mineralisations. There should not have been any gold there.

I contemplated the thought that some enterprising but otherwise dishonest person might have intentionally salted the hole by dropping some specimen stone down with the expectation that the high grade intersection would be reported to the Stock Exchange and react favourably on the share prices of the two listed public companies of the KMA consortium, Gold Mines of Kalgoorlie and Poseidon. As Chairman of KMA and GMK I was in an interesting position. Should I report the intersection and start a frenzy of activity on the exchange, or should I sit tight and say nothing. While I decided the next step I bound Jim Cleghorn to secrecy. He was the only person who knew about the sample, having collected it from the drill rig, and wisely had told no other person than me.

Besides being convinced in my own mind that it did not represent a significant discovery I recalled an incident that had occurred at the WMC iron ore operation at Koolanooka Hills in the late 1960s. One of the employees drilling for water a few miles east of the mine, salted some of the cuttings with gold as a practical joke and reported it to the mine manager who after inspection of the cuttings reported it directly to Melbourne Office. He had been generous with the gold he had added to the sample and Melbourne Office were just on the verge of releasing a report to the Stock Exchange of the high grade near surface (one hundred foot depth) gold intersection when the driller sheepishly admitted to taking a rise out of his boss. After careful thought I decided that whatever the circumstances behind the Mount Charlotte high grade intersection, geologically it was wrong and there would be no substance to it. It remained confidential to Jim Cleghorn and me.

The diamond hole made the expected intersection of the Mount Charlotte ore body at a depth of three thousand eight hundred feet and confirmed that the mineralisation persisted to that depth at somewhat similar grade to the unfaulted section and immediate action was taken to locate the optimal position for a new main shaft of six and a half metre diameter and to commence sinking.

About two years later when the associated development work had slackened in intensity a crosscut was driven five hundred feet out to the surveyed position of the high grade intersection. It was found to be a patch of quartz from which about a truck load of high grade stone was recovered. Further development around it failed to discover any more ore. The Stock Exchange may be critical of this behaviour but I am sure I took the proper action.

Had the management of Kalgoorlie Lake View been able accurately to assess the amount of gold locked up in their two old treatment plants it is quite probable that by recovering it, it would have been possible to have got through without taking in another entity. When the Lake View and Star plant was closed down and a systematic clean up of the plant was undertaken, commencing with the obvious gold traps, gold in an amount of approximately fifteen thousand ounces was very quickly recovered. One two-inch diameter galvanised pipe in the gold room was found to contain two thousand ounces of gold in the form of amalgam! Gold to a lesser amount was recovered from GMK's Oroya plant where the highlight was the recovery of one thousand ounces of gold from zinc-gold slime that had been quietly accumulating in the Merrill tank since 1935. In total over thirty thousand ounces was recovered from the Lake View plant.

Much of this was spent on "off budget" items, rehabilitation work for which Homestake wouldn't approve the expenditure but which had to be done to stop the plant from falling apart. The gold won from cleaning out the bottom and chipping the sides of agitators that had been operating for thirty or forty years and were corroded to such an extent that their replacement was essential, would pay for the new agitators. Much of the plant had been renewed in this fashion when in 1979 unanimity was at last reached by the associated companies to make a proper job of rebuilding the plant. Also in 1979 dewatering the Fimiston leases was commenced.

I had been pressing for this for two years but could not get the approval of Homestake. Prior to shutting down the Fimiston leases an assessment of proved and probable ore in both the Eastern (GMK) leases and the Western (LVS) leases had indicated a total of five and a half million tonnes at a grade of eight and a half grams per tonne. Although the development paid for by the $440,000 State Government assistance had been directed towards bringing this ore into production it had not gone far enough and it was not until 1981/82 that sufficient ore was developed to justify commencing production.

Had the re-opening of the Fimiston leases been approved in 1977, as I recommended when the gold price started its upward move, KMA would have reaped the benefit of the high gold prices when they reached their peak in 1979/80. The companies that do well are those able to take advantage of a rising gold price. When it reaches a peak it generally does not last long and the more cautious companies that defer development until it appears ultra safe sell their production when the price begins its downward slide. Nevertheless, GMK's share of the profits in those halcyon years enabled significant changes to the company's financial structure to be made.

☆ ☆ ☆

During the barren years post 1967, WMC had been lending GMK money to pay its administration costs, stock exchange fees, share registry costs etc. These totalled approximately $700,000. WMC management had strong reservations about the short to medium term future of gold and were not prepared to increase WMC equity in GMK as a means of putting GMK in funds, nor were they prepared to subsidise GMK any longer. This was in part due to the wish that WMC not be seen as 'big brother' in Kalgoorlie where it already had a dominant role, nor in the event, perceived as unlikely but a possibility, that gold would recover, to lay itself open to the charge that it had taken the opportunity to benefit from the misfortune of an associated company. I took over the Chairmanship from Sir Lindesay in 1974.

In 1977 WMC withdrew its two representatives – (Sir) Wilfred Brookes and Roy Woodall from the Board and I replaced them with Ted Baltis and John Carroll to represent the general body of small shareholders. WMC relinquished management of GMK and transferred the registered office of the company to Ernst & Whinney from which John Howard was appointed Company Secretary. It is interesting to note that during this transfer of management responsibility by WMC, John Howard was told by the WMC financial administrator that it would be a short term appointment as it was most unlikely the company would survive.

The Board had either to develop a plan to make the company solvent and regain profitability or liquidate it. The $1.00 shares were then trading at eighteen cents. The slow steady rise in the gold price from late 1976 justified going to the shareholders. To set the scene, GMK reduced the value of its shares from $1.00 to twenty-five cents and wrote off past losses of $4.6 million. A one-for-three share issue raised $510,000 which was used to reduce the WMC loan and provide working capital. Although WMC did not take up its entitlement it agreed to waive interest on the loan which then amounted to $643,000. By August 1978 the outlook had improved sufficiently to make a private placement of eight hundred thousand shares at sixty-five cents which extinguished the outstanding debt leaving a small remainder to supplement working capital.

The gold price continued to rise converting an operating loss of $232,191 in 1978 into an operating profit of $2,276,706 in 1979 and GMK again entered the dividend lists. From that point on until I retired as Chairman in 1985, we were laughing all the way to the bank. In 1983 GMK's operating profit from KMA operations reached $11,361,570 from which $7,289,831 were paid in dividends.

From 1968 to 1978 the company had paid no dividends to its shareholders. Although the new nickel industry had become a major support for the town of Kalgoorlie, the part GMK played was of equal importance. During those ten unprofitable years the combined reserves of the Golden Mile and Mount Charlotte were

SIR LINDESAY CLARK

67 7556

459 COLLINS STREET,
P.O. BOX 860 K,
MELBOURNE, VIC. 3001

December
19th
1975

L. C. Brodie-Hall, Esq.,
Western Mining Corporation Limited,
P.O. Box 80,
Belmont, W.A. 6104.

23 DEC REC'D

Dear Brodie,

Hearty congratulations on your achievement in getting Homestake to join with Lake View.

I would like the congratulations to extend to your career in the West in which you have made such a notable position for yourself entirely by your own efforts for many years. I am sure the people of Kalgoorlie, if they really knew what you had brought about for them, would echo these views.

Kindest regards to Dorothy and you and best wishes for Christmas and the New Year,

Yours ever
Lindesay

reduced by over nine million tons and nearly one and a half million ounces. By far the greatest part of the burden was carried by GMK before it amalgamated with Poseidon and subsequently Homestake. In total it ran its ore reserves down by five million tons representing about one million ounces. It is gratifying to record that when the crisis was over the very large operating profits generated annually by KMA paid for the construction of new plant, re-opening Fimiston mines and sinking the new Cassidy Shaft to a depth of twelve hundred metres at Mount Charlotte.

The shareholders got their reward. Before the investing public went mad and lit the fuse connecting to the time bomb that exploded on 19 October 1987, GMK shares were trading at $12.50.

I didn't participate in that bonanza.

In 1974 when I was appointed Chairman I decided I would cover myself against possible accusations of trading on insider knowledge by not holding any shares. When the shares commenced their dizzy rise in 1986/87 and were approaching $15 a share, my wife Jean had a small number of shares that she had held for many years, I advised her to sell, on my assessment that $8 a share was their proper value. Jean hung on until they reached $20 a share when she sold. They continued to rise and went through $30 and then $40 during which time she brought to my attention from time to time that I was a poor judge of the market. What I didn't know, and Jean didn't disclose to me until after, was that she had only sold one half of her shares. She eventually disposed of the balance for $49 a share. I am glad she had not told me, as I would have certainly done my best to persuade her to sell long before they reached their peak.

My reward came in other ways – a congratulatory letter in December 1975 from Lindesay Clark who, as may be imagined, has a special place in my affections.

Many years later the Councils of the two towns which are only five kilometres apart and had been in varying degrees of confrontation since their formation at the end of the last century, decided on 19 March 1989 to bury the hatchet and amalgamate into the City of Kalgoorlie-Boulder. The proclamation of the new City was marked by a celebration in the Town Hall at which the award of Freeman of the City was conferred upon me. I felt greatly honoured and deeply touched. Although I was born in London I somehow regard Kalgoorlie as my home town.

Chapter 15

It was a very black day in my life when Dorothy returned from a visit to the doctor to tell me that some symptoms that had been worrying her for some time had been diagnosed as cancer of the lower bowel for which she required an immediate operation. The specialist surgeon was not pleased with the result as post-operational pathological tests indicated that the cancer was present in the lymphatic glands and that the likely result would be that it would invade the liver and that her life expectancy would be short.

Dorothy adjusted to this very quickly. She was a strong spirited person and quickly decided that she would make the most of whatever life was left to her. Whereas I am a total duffer at all types of sport Dorothy was well above average and much of her social life was built around golf and tennis. She also played a good game of bridge. These were all things that fitted in with her fun-loving, gregarious nature. Whatever was missed from her life in these matters while she was bringing up the family and helping me with my career she tried to make up for in the ensuing twelve months while waiting for the next development. Confirmation of our worst fears came after about ten months with the pronouncement by the specialist that it was terminal cancer of the liver and that she had only about six months of life left.

Because of our different life styles we hadn't spent much social time together. I am so hopeless at sport that I concentrated my spare time on gardening and for evening relaxation classical music or reading.

At that late stage of our marriage (Dorothy was 59 and I was 67) we tried to catch up by going everywhere together which generally meant her accompanying me on my business trips. A few months later, towards the end of November she had to take to her bed. It was fortunate that my various board duties allowed me to transfer my work from office to home and look after her personally. However I had to go to Melbourne early in December to a WMC board meeting and left her in the care of our elder daughter for the four days I was away, during which time she completely dropped her bundle. When I returned it took me days to get her spirit up again but she started to look forward with some enthusiasm to having all the grandchildren home for Christmas, an event for which I bought the presents which Dorothy was able to parcel up and present to her five children, their spouses and her twelve grandchildren.

She felt so well after this that she asserted that she was going to hang in for quite a long while and that it was not fair to me to have to stay home to nurse her. I should engage a nurse and go back to work. I told her that the earliest date at which I would have to leave to attend the next board meeting in Melbourne was 28 January 1978 and that anyway I intended to excuse myself from attending it. Dorothy must have firmly fixed that date in her mind. Over the ensuing weeks her condition progressively deteriorated and with the strain of watching her waste away, added to attending to her needs around the clock, I had to seek help from those of her friends who had nursing qualifications. For ten days one or another would come in for twenty-four or forty-eight hours which allowed me to get some sleep at night. Every day during the final week, whichever one it was that was helping out would say in the morning "she won't last today out"! She eventually died on January 28th. I am sure that Dorothy had that date firmly fixed in her mind as the day that I should go to Melbourne, that she would hang on until that date and then depart and cease being a trouble.

The Brodie-Hall family, May 1977. Charles' Wedding Day
L to R: Lesley (seated), Brodie, Robert, Charlie, Phillip, Dorothy (seated), Kate (seated).

Chapter 16

John Howard, the Secretary of Gold Mines of Kalgoorlie, rang me from Melbourne one morning in 1980 in a state of great excitement. He had received a telephone call from Lynn Evans, one of the Principals of Austral Development Ltd, GMK's agents in London, who were also I understand greatly excited. Lynn had been visited by a chap who claimed to have been one of a party (of crooks) who had a plan to lift a large quantity of stolen gold out of Kalgoorlie. He had been dumped from the gang and hoped by thwarting their plans to be rewarded with a percentage of the gold if it were recovered by the use of the detailed information which he had of the gang's plans and which he was prepared to sell. The amount of gold he stated as being the subject of the lift stretched my credulity to the limit. He quoted a figure of two tons, say, sixty-four thousand ounces. The gold price at that time was $650 an ounce.

Gold stealing has always been rife in the goldfields, the general attitude of the majority of people in the industry being, if you can get away with it you are a mug not to. There are people prepared to buy it – at bargain prices – as specimen stone, amalgam or any other form in which it can be nicked. There were several illicit gold buyers living and carrying on their business in Kalgoorlie.

The story that this chap told – his name was Johanssen – described the gold as being an agglomeration of smelted gold, tael bars, two four hundred ounce bullion bars and assorted items of carat jewelry which would lead to the assumption that some of it had been obtained by direct robbery.

To combat gold stealing and illicit dealing in gold, the Chamber of Mines maintains a special squad of police in Kalgoorlie, the Gold Stealing Detection Staff. Their presence on the goldfields is more of a deterrent than anything else although occasionally someone was caught and brought before the magistrate who, in the eyes of industry executives, took the same relaxed view as the general community and awarded a penalty comparable to a slap on the wrist.

Fred Douglas, a retired Detective Inspector of the Gold Stealing Detection squad, was a friend of mine. He was a scrupulously honest man and in his day had been tireless and successful in the prosecution of stealers and dealers. He told me that most of his arrests, made as a result of tip-offs from informers, were made in the bush at the site where the gold buyer had a small plant for processing the high grade ore into bullion. To make an arrest required constant surveillance of the site twenty-four hours a day, sometimes for weeks on end.

Before either dismissing this news from London as a fairy story or getting wildly excited about it I decided to consult Freddie. To my surprise he didn't bat an eyelid at the mention of a cache of sixty-four thousand ounces. At that time he lived at least part of his time at Esperance, a port for the goldfields and a holiday resort for goldfields communities. He told me there was a lot of talk at the Port about gold from Norseman in significant quantities changing hands and expressed the opinion that if the gold had been accumulating for some years waiting for a rise in the gold price and the virgin gold was supplemented by stolen bullion and stolen jewelry there could very well be two tons. I digested that and having accepted his views as valuable asked him if he would investigate it further, go to England, talk to Johanssen and if convinced of the authenticity of the story negotiate a deal, offering as a reward a significant percentage of all gold recovered as a result of the information imparted. He was wise enough to decline and point me in the right direction. "This is a job for the CIB," he said, "If I got involved in it there would be great resentment in the force and you would probably find they would go out of their way to make things difficult for me. Go through the CIB."

So I rang the Commissioner who sent a Detective Sergeant around. Before sending him off to London where I had arranged through Austral Development to set up a meeting between the Sergeant, themselves and Johanssen and acting in my capacity as Chairman of Gold Mines of Kalgoorlie (Aust) Ltd I gave him written authority to enter into a legally binding agreement to pay Johanssen twenty per cent of the market value of all gold which was recovered as a direct result of his information. This was a bit cheeky because if the loot was recovered intact, the entire two tons of it, it was probable that the bullion bars would have been identified and returned to the rightful owners, neither could Gold Mines of Kalgoorlie have established ownership of the apparently virgin rough smelted gold to the exclusion of other companies still operating.

To this point I had told no other person than the Detective Sergeant and had assured myself that John Howard would maintain secrecy. I felt that I would benefit from third person advice in this matter which could call down criticism on my head from other senior members of the gold mining industry so I discussed it with my old friend and mentor Sir Lindesay Clark. He confirmed my opinion that the fewer people who knew about it the better. The Detective Sergeant went off to England and nothing further was heard about it until his return about ten days later.

Johanssen, who incidentally had an air pilots licence, lived in Wimbledon and had an associate (or accomplice) with the most surprising name of Sykes. This caused me some astonishment. I thought that any member of the criminal classes would have been more protective of his anonymity than to have retained a name so evocative of criminal activity as Sykes. Whether his first name was Bill, I don't know but for me at least he was Bill Sykes. After the parties had a preliminary meeting, looked each other over and decided it was reasonable to proceed to the next step, Johanssen and Sykes were represented at the next and subsequent meeting by a legal practitioner. After the deal was signed Johanssen spilled the beans.

The basic plan was to air-lift the gold out of Australia and sell it in the Middle East. To achieve this an air transport company had been formed and registered in Nigeria. It was proposed to use five DC3 planes ex the Royal New Zealand Air Force for which an option to purchase had been entered into with the New Zealand Government. The expiry date of the option was August 31st 1980 and as it was the end of July there was some urgency to close the deal. The planes would fly to Sydney across Australia picking up the gold en route and thence to Bahrain where the gold would be disposed of before flying the planes on to Nigeria.

At this stage the Detective Sergeant's superior officer, Detective Inspector Gould, apparently convinced by the authenticity of the documentation of the registration of the air transport company and his Sergeant's report, decided that he should take over and continue the investigation in New Zealand.

The planes were located and identified. The option for purchase was verified. Five RNZAF pilots, veterans from the second world war had been engaged to fly them to their destination.

The fly-out was to be initiated by a Grand Flyover of Auckland and the route would be Auckland, Sydney, Alice Springs, Broome, Rangoon, Colombo, Bombay, Bahrain and then across Africa. The gold would be transferred from wherever it was stored in the Kalgoorlie district by four-wheel drive land cruiser type vehicles to Broome where the new air transport company executives, i.e. the chief crooks, and the pilots were to be entertained at a dinner hosted by the President and Councillors of the Broome Shire. One key man would remain to supervise refuelling the planes and the transfer of the bullion and its stowage in the DC3's. So far, so good.

At this time a severe shortage of aviation spirit was looming. WMC had at that time an exploration party working in the Northern Territory, the work being under the direction of Richard Mazzucchelli. An important part of the work planned was an airborne geophysical survey. Richard was having trouble getting this lined up due to an impending national shortage of aviation spirit. He called me by telephone to ask if I could pull some strings to ensure the supply of aviation spirit to his geophysical contractor. I was not successful. I mention this because the aviation spirit shortage which extended to New Zealand may have been the reason why this elaborate plan was not implemented. The option expired and was not renewed. Why was it not exercised before the expiry date? Were the financial backers unable to produce the purchase price? Was it the aviation fuel shortage? Or was it that the crooks had become aware that their plan was blown? The unanswered question is what became of the gold?

Disposing of large amounts of gold through legitimate channels by people who had gained it by illegal methods, such as stealing or receiving stolen goods, presented a problem as the gold had to be identified with the mineral lease from which it was mined. This increased the risk that those endeavouring to sell it through the normal banking system would be found out. Hence the attempt to smuggle it out of the country.

There was another way which may (or may not) have been used.

During the 1980s a very large quantity of gold was sold in nugget form having been found by the use of metal detectors. It would have been no great problem for illegal operators to have melted the gold and poured it at random on to an irregular surface of sand, thereby simulating the appearance of nuggets. It could then be sold legitimately in small quantities. To me it is a credible explanation. I can think of no other easy alternative.

☆ ☆ ☆

There are very few people who go through life without experiencing a lucky break. Some of my friends, usually the more successful ones, say "You make your own luck," attributing all of their success entirely to their own efforts. Well, they may be right.

I had a great stroke of luck during the 1970s. It originated in 1973 and culminated in 1980 and many occurrences contributed to it. In my view it was pure luck.

Jim Lissiman, an old friend and ex Western Mining colleague, called me by telephone one day to ask if I could be interested in becoming a member of a syndicate he was putting together to re-open the old Nannine Gold Mine at Nannine in the Murchison. Going on its past production history, he thought it had a good chance of being brought back into profitable production. I had not heard of the mine before and I asked him to send me any information he had on it and on some associated leases that were part of the project. The main document was a report prepared in 1945 by Col Adams, one of the staff of the Western Australian Department of Mines. At the time of publishing the report he was District Inspector of Mines at Cue, another old mining town about eighty kilometres south of Nannine.

It was a good report which collated such records of exploration and production as there were dating back to the discovery date and including a report by the de Bernales group from when it had been active in the 1930's. The Nannine was the first payable mine discovered in the Murchison Goldfield and its production of about one ton of gold at around one ounce to the ton was very encouraging as the mine had only been developed to the sixty-one metre level. The de Bernales diamond drilling had indicated good grade ore continuing at least to one hundred and fifty metres.

After closely examining Adams' report I decided that it would be a very good punt. I say punt because in my experience there are many mines with good production records which according to folk history closed down "because the water beat them," when in most cases closure resulted because the ore became uneconomic at the sulphide zone which is generally around the water level. My gambling instinct is practically non-existent. I don't follow the horses, never go to a casino and seldom buy a lotteries ticket. I speculate on the Stock Exchange but usually after close research into the proposed investment. I have for long held the view that whatever I get I would have to work for. The Syndicate proved me wrong. After close self-analysis I have come to the conclusion that I joined the syndicate because I was an incorrigible gold buff!

It was to be a small syndicate of ten units. The leases were held by a well known Murchison prospector, Tom Parkinson, a young fellow of seventy or so, and he would receive two units for them. Jim Lissiman, the initiator of the project, two units, Lou Rinaldi of Meekatharra, the initial site manager two units and one unit each were offered to and taken up by three other hopefuls

Nannine, Longyear Steam Drill, 1936.

and me. The initial contribution was $1000, equivalent in today's money to $6,000. The gold price in 1973 ranged between $66 an ounce in January to $100 in December with a two months spike of $125 in June and July.

The first task was to get the water out of the mine and thc necessary equipment (pump, engine and pipe) was leased from Tom Parkinson and installed on the Nannine main shaft. The water was quite saline, ten per cent total dissolved solids by weight and highly corrosive; by comparison sea water is three and a half per cent. Before the water was lowered to the sixty-one metre level the pump and rising main were badly corroded and the contribution of the unit holders had risen to $5,000. The gold price had been rising steadily since the formation of the syndicate and when samples of the ore shoots on the newly exposed level confirmed the ore grade and their continuity we were all very pleased. Regrettably the gold price, having reached $195 an ounce in December 1974, started to drift downwards. The kitty was almost empty and when the gold price started to decline so did the enthusiasm of the unit holders when asked to put in more money. The pumping equipment was in very poor condition and the water slowly rose again to its original level. The mine stayed flooded until 1979. The gold price which had bottomed at $103 an ounce in August 1976 started to rise steadily and when it went over $400 an ounce in 1979 the unit holders were willing to start contributing again. Better still, other investors were asking to come in at very substantial premiums on the initial unit price.

There had been an occurrence during the four year shut-down period which was a strong deterrent to feeding more money in. On the night of 5th September 1975 a case of gelignite and some detonators were stolen from the explosives magazine in Meekatharra. The following night the collar of the Nannine main shaft, where the pumping equipment had been installed and which was our means of access to the sixty-one metre level, was blown up and the top seventeen metres of timber destroyed. We assumed that this was done in the hope that in the face of the worsening gold price and the pumping problems, the syndicate would lose heart and relinquish the leases. Nothing doing there. We dobbed in more money when required to keep the leases in good standing and our applications for exemption from the labour covenants were granted.

With the injection of fresh capital new pumping equipment was purchased, this time manufactured with stainless steel, and installed on another shaft two hundred and fifty metres north of the main shaft originally used. The water was got out, the shaft sunk to the seventy-eight metre horizon and eighty metres of level development completed which picked up two of the ore shoots confirming the expected size and grade. A diamond drilling programme of five holes returned two intersections of quartz reef at the one hundred and fifty metre level of approximately one ounce per ton. The others were a bit off target but encountered reef with low grade gold. Two good intersections out of five is a satisfactory result when testing ore containing spotty free gold and, taken in conjunctions with the good results of the seventy-eight metre level and the earlier drilling, consideration was given to forming a public company. Following satisfactory reports from various consultants, Hartley Poynton & Co were retained to prepare a prospectus for the issue which they agreed to underwrite. At this point the Nannine Syndicate had been converted to the Nannine Unit Trust comprising sixty-one members who between them had spent nearly $1,000,000 on development and various capital items. The vendors' consideration was 9,045,360 ordinary shares of twenty-five cents each. A total of 4,522,650 options were granted to the vendors.

These negotiations were well in hand when the Unit Trust received an offer from Alan Bond of $9 a unit for each and every unit of 753,780 units on issue. This offer was so very much better for the shareholders than the issue of shares in the new company that I was amazed when Jim Lissiman visited me to tell me that there was sufficient opposition to Bond's offer to cast doubt on whether it would receive the plus seventy-five per cent acceptance necessary to approve the offer. He asked me if I would address the meeting and describe to them the advantages of the Bond cash offer over the alternative issue of shares in a new company.

This I did along the following lines. A mineral boom of such magnitude that it would absorb all of the quality miners available had been forecast for Australia. A mine such as ours would require top quality employees and would suffer in competition. A small rich mine is very vulnerable to gold stealing and the cost of providing adequate security surveillance is high. The gold price is very unpredictable and a substantial drop in the price would eliminate the profit margin. Being vendor shares they would be in escrow for twelve months, during which time there could be a dramatic fall in the price. These and other considerations I explained to a very well attended meeting of unit holders. When the vote was taken the motion fell short of the necessary seventy-five per cent by about four per cent. Most of us were sitting there feeling glum when Terry O'Connor came up with the suggestion that we should tell Bond that the offer of $9 was considered to be unacceptably low and if he was prepared to lift his price, a figure of $11 a share would be acceptable. I thought his reaction would be,"Well if you want to bargain over the price I withdraw my offer of $9 but I'll give you $8." Jim Lissiman and Antun Triglavcanin saw him immediately after the meeting and Bond's reply was, "I won't give you $11 but I might give you $10. I'll call you in the morning after I have thought it over." He was as good as his word. Bright and early the next morning we had an amended offer of $10 a share with acceptance or refusal within twenty-four hours.

Another special meeting of unit holders was called. In the meantime some fairly intensive lobbying had been going on with some of the dissident shareholders. The main supporters of the proposed new company had been Hill Minerals and Target Paru two companies which between them held seventeen per cent. It was a very tense meeting and when it came to the vote the air was electric. Obviously some of the shareholders had not heard the old adage that 'a bird in the hand is worth two in the bush': With some people avarice knows no bounds. A Mr Schmidt, from a famous broking firm in New York, was in Perth at the time and during a television interview had stated with great conviction that the gold price would reach US$1,000 in the immediate future. It was at that time about US$800 an ounce. I am firm on that statement, and from the recesses of my memory I recall that two elderly ladies – also American – who were something in the financial world had thrown caution to the winds and were predicting $1,500 an ounce.

When the votes were counted the result was seventy-five plus something – it exists in my memory as seventy-five and one fifth per cent – in favour of accepting the Bond offer, the main voters against being Hill Minerals and Target Paru. There were some other temporarily unhappy unit holders who believed the price would continue to rise and that if the company had been formed the share price would have rocketed to a level which would have far surpassed Bond's offer. The legal details of transfer were transacted with commendable speed and within thirty days (19/12/80), a cheque for a little over $7,500,000 was paid to the account of the Nannine Unit Trust. The management lost no time in distributing it. My original ten per cent share had become watered down to seven and a half per cent and my share of the booty was about $525,000, at a cost of just under $10,000.

On Christmas day I was in the happy position of being able to give the eight children of my extended family - three of Jean's children and five of mine - $10,000 each and $20,000 to Douglas, my elder brother who was staying with us at the time, and to have a handy sum to put in the bank against a rainy day.

The deal was transacted at the peak of the gold price. It almost appeared as if our deal with Alan Bond had been the trigger for it to start falling. It bottomed out in June 1982 at US$295 or about A$260.

A friend of mine once drew my attention to the fact that no one rings a bell when share prices have reached their peak. The same observation applies to the gold price.

If all of those contributing situations and occurrences don't add up to a very large slice of luck, then I don't know what luck is.

Two of my friends who were in the original Syndicate put their money into developing a vineyard in the Margaret River area and now produce a range of wine sold under the name of Fermoy Estate; very good wine, you can taste the gold in it! Another put his into property which takes care of his future very nicely. I put some of mine into a little farm (ninety-eight acres) at Harvey where my wife has a small herd of cattle and raises baby beef. I was very tempted to call the property "Bond's Folly" but settled on the less provocative "Nannine Farm". The same friend who gave me the advice about the absence of a bell ringer when share prices are peaking also drew my attention to the complete absence of pockets in shrouds. Two years later I was raising money for an Education Trust Fund for the Western Australian School of Mines and decided I could not do much better than to give a good start to the fund with some of this "lucky money". My $135,000 donation is still the largest, outdoing the next nearest, a mining company, by $60,000.

L to R: Jean's daughters Ann with Julia's daughter Amanda, myself and Julia.

L to R: Charlie, Phillip, Robert, myself at the farm, Christmas 1991.

Chapter 17

My entry into the School of Mines was the turning point of my career or rather, I should say, opened the way for a career. Had I continued on the path I had followed since my arrival in Australia there would not have been a career. Acknowledging that, I have been conscious of the debt I owe the School and have maintained ongoing active interest in it.

For many years after it was inaugurated, first in Coolgardie in 1902, as an interim measure whilst the School premises were being built in Kalgoorlie to which it was transferred on completion of building in 1903, its curriculum was quite basic and the qualification it awarded was an Associate Diploma. In its early years it earned a high reputation but I think that this was due more to the calibre of its graduates and the practical knowledge they had of the industry than to the course structure and content.

From inauguration until 1969 the School was under the direction and management of the Western Australian Department of Mines, an association which assured its continuity during periods of industry down-turn.

During and after the Great War (1914-1918), the Kalgoorlie mines became less profitable due to lower ore grades and a high inflation and the School enrolments declined as a result. During the 1920's the industry and the School were at a low point, the Great Depression of 1929 et seq had been holding down the international economy and it was not until the United States President reduced the value of the US dollar in terms of gold by sixty per cent that growth returned.

When I enrolled in February 1936, School enrolments totalled seven hundred and fifty or thereabouts. They fell away during the Second World War and picked up again with the introduction of the Commonwealth Rehabilitation Training Scheme.

The School traditionally offered duplicated courses for the benefit of students working rostered shifts which made it possible for a shift worker to attend lectures alternating from day classes to evening classes without breaking continuity of instruction.

The courses included a year of preparatory study in Mathematics, Chemistry, Physics, Geology and Mechanical Drawing. Although I had retained some knowledge of the first three of these from my early education in London, I decided to lay a good foundation to my studies and signed up for the full preparatory year.

I had a brief but sharp altercation with the Registrar, who almost refused to enrol me for the five subjects as they were considered to constitute a year's full time study. I was employed as a continuous rostered shift worker – as a treatment plant operator I was entitled to two days off in a calendar month! However, my blandishments won him over to the extent that he enrolled me for the full course whilst forecasting that I would fail and would waste my tuition fees.

The fees were quite modest. A subject that entailed three hours of course work per week cost fifteen shillings ($1.50) a term. Five subjects probably aggregated twelve hours per week at a cost of £3 ($6) per term. My average daily wage during the first two years was about fifteen shillings ($1.50).

That first term was pure pandemonium, Student ages ranged from fifteen to fifty years with the younger ages in the majority. Class numbers in the preparatory subjects were about one hundred and twenty with most attending the evening session. The picture of Charles Warman, who subsequently founded Warman International, a very successful company manufacturing equipment which was sold world wide to the metallurgical arm of the mining industry, and who was at that time lecturing in Engineering Drawing, lives in my memory still. Naturally a gentle and soft spoken chap, his efforts to gain control over his class which was being kept in a state of near riot by the "just left school" contingent, would have broken any disciplinarians heart. His rather plaintive plea of, "Give me a go you chaps, give me a go," was inaudible beyond the first row of desks.

Order was usually established at the beginning of the second term (there were three terms in the year) by the simple expedient of thinning the students out at the first term exam. "Behold the rabble," said Sam Edelmann to me with a sweep of his arm towards the crowd of kids supposed to be studying Prep Chemistry but devoting most of their time to generating horrible stinks. "Just wait until the first term exams. I'll thin them out." I think he finished up the year with fifteen out of well over a hundred.

It was a year well spent for me. Despite the distraction, I waltzed through the year with good results and found myself in the following year with a number of chaps who couldn't face up to taking the preparatory subjects and then failed the first year subjects and had to repeat.

It was not without its humorous events. Edgar Elvey (a chap about my age who later became the General Manager and Director of the Great Boulder Mine), keen to get the best of both worlds, when working on an experiment which required a solution to be evaporated to dryness, would slip over to the Kalgoorlie Club (just across the road from the School and of which he was a member) for a quick drink. His assessment of the time required for the evaporation was on one occasion at fault and the residue in the beaker was giving off copious yellow fumes. This had attracted Sam Edelmann's attention and he was on the spot when Edgar returned. "And what do you think you are doing with that, Mr Elvey?" he asked.

"I have just discovered a new element," Edgar replied, "and am in the process of determining its atomic weight."

"Very interesting," said Sam, "And what are you going to call it?"

"Elvite, naturally, Sir." It was incidents like these that kept the lecturers, who weren't much older than some of the older students, from going bonkers.

There was one very lovable old gentleman, Harry Charlton Dod, who lectured in the more advanced subjects of Engineering, Chemistry II, and Metallurgy II. He knew his subjects very well but he was eccentric and took great pleasure in concocting a variety of pharmaceutical products which he tried out on any student game enough to let him. He also professed to be psychic and laid himself open to the accusation that he used super sensory powers to assess the quality of the examination papers submitted by his students. I complained on one occasion to Eddie Johns the lecturer in Engineering about the long delay between the posting of his examination results and those of Harry Dod. "I examine every exam paper in detail," he exploded. "Old Dod holds them up against his forehead for a second or so and then says that's a good one or that's hopeless and awards a mark accordingly." I choose not to credit that story, I always got very good marks from Harry Dod and I am sure that I earned them.

To those of us who maintained a regular regimen of study and saw the courses out until the end, the School became very much a club. Not that it had a suitable room in which to forgather but it resulted in forming close ties with a group of chaps with similar aspirations and I suppose personalities. Most of my School of Mines contemporaries have dropped off but have left with me a library of happy memories.

After graduating there was a break in my direct involvement in the School's activities between 1948 and 1968, years when I was making my way with Western Mining Corporation at Norseman, Melbourne and Bullfinch. From 1956 to 1964 when I was resident in Kalgoorlie, the years were crammed with new projects - bauxite/alumina, iron ore, talc and Mount Charlotte – which involved a lot of travelling and left no time for outside interests. When I moved to Perth in 1964 and Arvi Parbo went to Kalgoorlie as my deputy, I had more free time.

In 1968 Sir David Brand, State Premier, established a Commission under the chairmanship of the Chief Justice, Sir Lawrence Jackson, to enquire into and report on tertiary education in WA. In considering the future of the WA School of Mines he put forward three options:

1. To remain under the direction of the Mines Department;
2. To become an autonomous college;
3. To become a school within the Western Australian Institute of Technology (WAIT).

The decision rested with the Chamber of Mines as it represented virtually the entire mining industry in and around Kalgoorlie. The Chamber told the Commission it wished the School to become an autonomous college answering to a panel of local governors on which no doubt the Chamber would have strong representation.

Although I was a Vice President of the Chamber at the time I was not present at the meeting and unaware that the Commissioner had been officially informed of the Chamber's decision to which I was totally opposed. When I heard about it I was very put out. An autonomous college would always have funding problems which would restrict its growth. Lacking affiliation with a large institution there would be no avenue for staff exchange or promotion through a large organisation and the quality of staff would suffer.

Fortunately I had a friend in Ian Wark, later Sir Ian, who was then the Commissioner of Tertiary Education in the Federal domain. He was closely aligned with the mining industry as he was a metallurgist by profession and had many years of service with CSIRO. He fully supported my opinion that the School of Mines should become a part of the WAIT and took sides on the issue. The Brand Commissioners were quite open on the matter but it took six months to bring the Chamber of Mines around to my view. The incorporation of the School of Mines into the WAIT was finally agreed at a dinner party at my house at Gooseberry Hill in 1969 at which all parties were represented and happy with the result.

From that time on my active interest increased. I could see that with the added prestige that affiliation with the WAIT would confer on the School we could look forward to a substantial increase in enrolments.

Accommodation had always been a handicap. Lacking a residential college, students had to find board and lodging wherever it could be had. I had personal experience with these conditions during my first three years of study at the school when my quarters were a corner of a back verandah at a private house that had been partitioned off with recycled filter cloth. The fact that my landlady and her husband were very kind to me did not compensate for the discomfort and lack of amenities. The meals at the eating houses that catered for working men left a lot to be desired.

These conditions prompted me to propose to the Executive Council of the Chamber of Mines that the Chamber should raise a fund to build a residential college. This proposal was greeted in stunned astonishment by the other members who obviously thought that I was crazy to think that it was possible to raise the amount of money required. Their response was to the effect that if I would like to take it on then go ahead but don't involve the Chamber in it.

I thought it could be done and got a preliminary quote of $400,000 for a kitchen, dining and administration complex for one hundred and fifty people and two dormitory halls each with twenty-five bed/study rooms. These were named the Moore and O'Connor Halls in honour of Dr Bertie Moore, a long serving Principal of the School, and Mr C. Y. O'Connor the great engineer who designed and supervised the construction of the Goldfields Water Supply Scheme, a most daring project, to take five million gallons per day of fresh water from the coast to Kalgoorlie via a thirty inch pipe three hundred and sixty miles in length. Conceived in 1896 and approved by Parliament the same year when the population of the State was not much more than 100,000 it was completed at a cost of £2,800,000 and water flowed into the terminal reservoir in 1903.

It seemed reasonable to me to get $100,000 out of the mining industry, $100,000 from the State Government and matching funds of $200,000 from Canberra. To put these figures into perspective the 1970/71 cost is equivalent to $1.875 million in today's value. I sounded out Ken (later Sir Kenneth) Townsing, Under Treasurer for the WA Government, who was feeling in an expansive mood that morning and told me that for such a good cause he could find the money which he felt sure the Government would approve. The relevant Minister in Canberra guaranteed the matching funds and the real test came in getting the money from the mining industry. It was my first attempt at fund-raising on that scale and realising that it would be easier to get the money in large amounts rather than from a host of small donations I targeted four major companies for $25,000 each ($117,000 in 1993 dollars).

I learned quite a bit from that exercise. I tackled WMC first. When I put it up to the Board, (Sir) Wilfred Brookes was vociferous in his objection to giving so much, even $10,000 would be too much! He even questioned whether it should be supported at all. When someone carries on like that with an air of great conviction, other people are only too eager to follow the lead. Lindesay would have known that I am sure, for as soon as Wilfred ran out of steam, although his face was still red from the intensity of his emotions, Lindesay spoke up. "The School of Mines is in our backyard, you know. Most of our staff are School of Mines graduates and they are very good. We must support them. We will give them fifty ($50,000)." Phew! What a relief! That gave me heart to get to Maurice Mawby, later (Sir Maurice), Chairman of CRA and Hammersley Iron from which two companies he produced $12,500 each. Alcoa generously gifted $25,000.

With these successes I tackled BHP who I thought, as the Big Australian, would willingly come to the party. Ian McLennan greeted me most graciously and after I had told my story commented on what a wonderful proposal it was and then asked, "What in the name of the devil made you think BHP would give you anything? BHP has got nothing in the West." I corrected him on that statement pointing out the several important businesses they either owned outright or joint ventured. His comment that my request would have to be considered by the donations committee but not to worry as he was its chairman, renewed my fast diminishing hopes. About one month later I received a note from BHP telling me that they were pleased to donate $5,000 to such a worthy cause. I wrote back and told them it was not enough and looked forward to receiving a further sum the following year. It was a nice thought!

But through WMC's money I had the $100,000. In the meantime the architects costings had risen to about $550,000 and I had to ask them to prune them back which they did and, despite the cost cutting, succeeded in winning an award for the design. Forbes & Fitzhardinge were the architects who assigned the project to Tony Brand. The rest of the money was scrounged up among the mining service companies and some of the graduates. One chap, Matt Fitzgerald, who was General Manager of the large copper mine in Botswana, indignantly refused to contribute on the grounds that he had made his own way in the mining industry despite the School of Mines! I have had all sorts of reasons put up to me why people or organisations should not support an appeal but I thought that took the cake.

We had been fortunate in obtaining the title to Commonwealth Park, a barren piece of ground sufficient for our needs in close proximity to the School and which would permit the expansion of the college to its full potential of one hundred and fifty resident students. Jean Verschuer, a professional landscape architect retained by WMC as their consultant, contributed her services at no cost and successfully persuaded a local earthmoving contractor Ted Jessup to bring in a few hundred cubic metres of good red Goldfields soil for free. Humes donated the paving slabs, the WAGR carried them for us and the Town of Kalgoorlie provided the labour to lay them. Trees were donated by the Forest Department Nursery from Dryandra, and the only cost incurred in the landscape component was the irrigation equipment. After some 'instant gardening' supplementary to the permanent planting which would take a year or two to develop, Sir Charles Court declared the Agricola College open on September 26th, 1971.

The choice of Agricola as the name for the college may seem rather odd. It was the pseudonym of a German named Bauer who wrote 'de re Metallica,' the first text book on mining methods and extractive processes in use in the middle ages. I guess he adopted the Latin name to give added credibility to his treatise. The Latin text was translated in to English by Herbert Hoover, mining engineer and ex President of the USA, and his wife. Hoover was active in the eastern goldfields of Western Australia in the early days of the field as West Australian manager of the English firm "Bewick Moreing", Consultants and Mine managers. In the middle eighties his grandson visited Kalgoorlie and presented a handsome edition of the book to Agricola College, where it is on display.

When WMC completed the construction of its nickel refinery at Kwinana in 1970 and were planning an opening ceremony, a costly event that would bring together people (including quite a few from Europe and North America) who had been consultants or suppliers of specialised equipment or had signed long term contracts for the Refinery products, I suggested to the Board that they might consider, in lieu, endowing a Chair of Environmental Science at one of the Universities. They liked the idea and it was adopted. The Opening was very low key, one ceremony on site, to which Government ministers and local consultants and suppliers were invited, and another at the University of Western Australia. In fact, by arrangement between the Universities, the Endowment went to Murdoch University.

Alcoa of Australia's Kwinana Alumina Refinery, on the shores of Cockburn Sound, Western Australia.

Kalgoorlie office of Western Mining Corporation in 1960 at 55 Macdonald Street. This home was used by Herbert Hoover (later President of the United States of America) during the period early this century when he was manager of the Sons of Gwalia goldmine.

On the occasion of the awarding of the Australasian Institute of Mining and Metallurgy Medal in 1976 by the President of the Institute, J. G. Symons.

When the construction of the nickel smelter was nearing completion in 1972 and thoughts were once again turning to an opening ceremony, I sensed an opportunity to divert the money it would cost to the enlargement of Agricola College and to make the new construction a memorial to Bill Morgan, the recently deceased Managing Director of Western Mining Corporation. Once again the Board endorsed the idea and contributed the total funds required to build another two-storey dormitory hall of twenty-five single rooms which was named the Hoover Hall. Hoover is probably best known in the annals of the Eastern Goldfields as General Manager of the Sons of Gwalia. After attaining an international reputation in his field he went on to become President of the United States of America but with the misfortune that he held the position during the Great Depression. Bill Morgan was commemorated in an amenities area including a squash court, music room and study rooms. I can fairly safely say that the cost of the new construction was vastly greater than even the most prestigious Smelter Opening that could have been contemplated. It was a few years before all the accommodation was utilised as enrolment of full-time students had slowed down. However, opportunity only knocks once and on this occasion the best use was made of it.

Sometime in 1978 I had a visit from Bob Searles, the Managing Director of Newmont Mining, the fortunate finders and developers of the Telfer gold deposit. A glance at the map will reveal Telfer's remote location. Bob had found that among his best staff at Telfer were those who had not gone on to university in the normal course of events on leaving high school but had gained a tertiary education at the School of Mines much later at a mature age. He proposed a scheme to provide scholarships at the School of Mines for such people to which Newmont would contribute $100,000 spread over five years in equal $20,000 moieties to get it up and running. I suppose that having originated the idea he thought it was in order to leave me to implement it. I was one hundred per cent behind it and agreed to develop it further. The support that I received was phenomenal. Newmont's donation set the standard for other firms to match. Within a short time I had five other companies contributing on similar terms.

The scholarship was structured to provide a married man with $7500 a year, which the Commissioner of Taxation kindly exempted from income tax. The school year was of forty weeks duration, the remaining twelve weeks could be split into four weeks holiday and eight weeks of vacation work which would bring in, for an experienced man, another $2,500 of annual income. This being below the taxation threshold aggregated a tax free income of $10,000 per year which enabled a scholarship awardee to bring his wife and children to Kalgoorlie, rent a house and maintain a normal home life while he worked like a beaver at school. A scholarship for a single man was $5,000 a year.

The criteria were that the student must be mature age, have evidence of having passed the leaving certificate, carry a recommendation from a previous employer and satisfy a panel of the Fund Trustees that he possessed the motivation and the personality to give reasonable hopes of a satisfactory conclusion. The history of the scheme which was named the Mining Industry Tertiary Education Scheme is impressive. Of forty-one grantees, thirty-five graduated in their chosen course. Some students required a fair amount of additional tutoring which was paid for out of the Fund but this was needed only in the first one or two years and generally for mathematics.

It was fine during the first few years when the scheme was working on the original pledges. As the years rolled by the effect of inflation had to be offset by adjusting the value of the scholarship by percentage increases equivalent to the CPI. In later years this became beyond the capacity of the Fund to maintain. It now stands at $17,500 and the CPI index was dropped in favour of an arbitrary sum decided annually. The number of scholarships that could be funded decreased until in 1993 none were awarded. There are so many mining companies now supporting scholarships to school leavers that it is doubtful if the scholarship for mature aged students will survive.

The School had become moribund for several years when the position of Principal became vacant in 1976 and a Welsh academic with a mining background, Dr Odwyn Jones, was appointed. Here was a man with fire in his belly and the intention of raising the courses offered at the School to an international standard. Unfortunately almost from the time of his appointment his ambitions in this direction were impeded by government intervention. A report on Tertiary Education in Western Australia by Professor Partridge recommended inter alia that many of the advanced subjects at the School of Mines should be transferred to the Perth campus of the parent body, the Western Australian Institute of Technology. The recommendation, made predominantly on economic grounds, was strongly resisted by the Chamber of Mines and the Kalgoorlie community. A committee formed to review the Partridge report reversed his 1977 recommendation and proposed in lieu a Federation of the Kalgoorlie-Boulder Technical College and the WA School of Mines where each would maintain its identity but by cooperative sharing of premises and staff would hope to reduce costs. Much work was done on this proposal under the Chairmanship of Jack McDermott and it was nearing the stage of official acceptance and implementation when out of the blue the Hon Peter Jones, MLA, Minister for Education, announced that the Federation concept had been scrapped and that the two institutions were to be merged into a Community College with the title The Western Australian School of Mines and Further Education (WASMAFE). A committee was formed forthwith (in December 1980) under the able chairmanship of Jack Manners and much good work was done in the six months preceding the recognition by one of the academic staff in July 1981 that if the WASMAFE was established the School of Mines would lose the College of Advanced Education status it enjoyed as a branch of the Western Australian Institute of Technology.

Several senior members of the academic staff had resigned during the previous years when no one could be certain of the future of the School and this latest development caused quite a furore among the staff, students and the Executive Council of the Chamber of Mines. In an effort to get the School on course again, the Chamber of Mines of WA (Inc), the students and staff requested that I be appointed Chairman of the WASMAFE committee. This was very bad news for Jack Manners who had been doing a sterling job in setting up conditions under which WASMAFE would operate. Not surprisingly he was in favour of the concept. I, on the other hand, was opposed to it for the same reasons that I had opposed the School becoming self governing. I therefore welcomed the opportunity to take over the chairmanship and get it on the right course again as a school within the Western Australian Institute of Technology. I don't think that this was what the Premier, Sir Charles Court, had in mind when, on appointing me to the position in September 1981, he said he thought I could get the situation resolved and WASMAFE established by the end of the year. The original committee of nine, drawn mainly from the local community, was expanded by the addition of six representatives from the mining industry. The committee met monthly for six months before the issue was decided by vote with eleven in favour and four against retaining the status quo, i.e. with WASM incorporated with the WAIT and the Technical College with TAFE.

Arising from this, in June 1982, the WA School of Mines was raised from being a School within WAIT and given Branch status with a much increased level of autonomy and the Technical School became Kalgoorlie College. It proved to be the best decision for the WA School of Mines. The immediate advantage was that Odwyn Jones felt secure for the first time in five years.

Self, Dame Leonie Kramer, Professor Odwyn Jones

The longer term gain was when the WAIT was given University status in 1987 to become Curtin University of Technology.

I was offered and accepted chairmanship of the Board of Management of the WA School of Mines and then ensued nine of the most rewarding years of my life in helping Odwyn to achieve his objective of raising the School to international standards.

By the time WAIT was translated into Curtin University of Technology in 1987, the School was offering three year Bachelor of Applied Science degrees in Mining Engineering, Metallurgy and Geology, four year Bachelor of Engineering degrees in Mining Engineering, Metallurgy and Geology, a Masters degree in Mining and Mineral Technology and a Doctoral degree in its own right in each of the above disciplines.

An early move on my part after being appointed Chairman of the Board of Management was to establish an Education Fund for the School. I had as a target amount one million dollars. To ease and simplify the work of fund raising I tried to get it from one person or organisation and thought Lang Hancock could be that person. At the first approach I thought that the idea of the Hancock Mining Education Trust had appealed to him. He certainly had the capacity to fund that amount as the Royalties he was earning from his contract with Hamersley Iron were very large. Friends close to him told me that if he didn't like what was

Kalgoorlie School of Mines, August 1989.

Sir Arvi Parbo, April 1988.

Sir Lawrence Brodie-Hall CMG
Gerard Darwin FRBS, FRSA
Bronze (life-size)
School of Mines, Kalgoorlie, W Australia

being put up to him the interviews were terminated after ten minutes or so. As he listened attentively and discussed the background to my request for a bit over an hour and then asked me to send him the proposal in summary form I thought that the idea of the Hancock Mining Education Trust Fund had grabbed him. He kept me on the hook for three or four months and then just before Christmas when I managed to get another appointment with him he told me that he had so many projects being worked up which if they came to fruition would require a lot of capital and that he just couldn't afford it. Regrettably he didn't agree to contribute any lesser amount and so it was back to the hard slog. I got significant donations from several mining companies but my contribution of $135,000 was almost double the next highest gift of $75,000. I failed in my objective to raise one million, only getting it close to $500,000. As at the 30th June 1993 the corpus of the fund stood at $553,394. By March 1993 $743,646 had been distributed to the School's Departments from income earned for the purchase of approved equipment or to pay for study tours.

During 1984 Odwyn and I had been discussing the idea that the School should become more involved in research for the mining industry and had in mind a model similar to the Julius Kruttshnitt Centre established at Queensland University. The JK Centre was actively and successfully soliciting research projects in Western Australia and we thought we should endeavour to keep some of it in Western Australia.

In 1985 we decided to go ahead and I got it away to a good start with a personal donation of $100,000. It was probably during a period when the share market was up and on paper I seemed to be able to afford it. The situation changed dramatically after October 1987 but that bears out my belief that the time to be generous is when you have got the money. Anyway, initially the academic staff seemed to approve of the idea and they insisted that the new establishment be called the Brodie-Hall Research and Consultancy Centre. Later when they contemplated the inroads the Centre could make in the supplementary income they earned from private consultancy they changed their stance. Academics are woefully underpaid and I could see their point of view. To help them maintain this additional money while also helping the fledgling Centre I proposed that their private consultancy work should be channelled through the Centre which would retain fifteen per cent of gross revenue to cover use of equipment and administrative services which also included indemnity insurance provided by the University. I was both surprised and disappointed when they rejected the offer having thought they personally would like to support the School by passing their work officially through the Centre and getting tangible value in return even though they would then receive payment through official channels. I guess I didn't take into account the frailty of human nature. In the face of this resistance none of the staff were prepared to take up the position of Director of the Centre and Odwyn took it on.

Gold prices which had been steadily rising since 1983 peaked in 1987, reaching a high of $707 an ounce in October and November for an annual average price of $639 an ounce. Although the price declined progressively in 1988, the average price for the year was still quite good at $563 an ounce and many of the larger companies had sold their gold forward in 1987 when gold was at its peak. In view of the high profits most gold mining companies were enjoying it seemed to be a good time to take the hat around.

Odwyn was very keen to establish a Chair in Mineral Economics and Mining Engineering. The School needed another lecturer in Mine Surveying and for years Odwyn had been endeavouring to get enough money for an Instrom rock testing machine for the Brodie-Hall Research and Consultancy Centre. Enrolments were also rising and additional accommodation was needed and in the back of our minds we had a long held hope that some time we could get enough money to build a Conference Centre.

I have never used the services of professional fund raisers nor have I ever had a committee to assist in canvassing potential donors, preferring to do it on my own with secretarial assistance from my secretary or my wife. The response to my appeal was staggering, reaching a total of four million one hundred thousand dollars ($4,100,000) generally pledged over a five year period. One private donor who asked to remain anonymous gave $250,000 in a lump sum towards the Chair of Mineral Economics to which Metana Minerals had gifted $250,000. Four other companies gave $250,000 and there were many significant donations of lesser amounts.

At Agricola, forty self servicing units were built for students mainly in years three and four of their studies, it being the developing trend in student accommodation. It was always the practice to house year one students in the dormitory houses where they could be adequately supervised. My investments were in a fairly sound position and I was able to donate $100,000 to this work program also.

A major achievement at that time was getting funds for a Conference Centre. I put this proposal in a letter to Sir Arvi Parbo,

Sir Ninian Stephens awarding the honorary life membership of the Australasian Institute of Mining and Metallurgy, 1988.

Chairman of Western Mining Corporation. In retrospect, the cost I attributed to it was rather low and Arvi recognised it as such. It wasn't a guess on my part but a rather hasty estimate that had been made by a reputable builder. Arvi's reply, while casting strong doubts on the figure quoted, left the way open for me to put up a more accurate figure. When the detailed estimate was prepared it came in at $1,300,000, several hundred thousand higher than my first figure.

In the expectation that at that price my request would be refused, I had an estimate prepared for the construction of a Research Laboratory which came in at a bit over $700,000. Armed with these estimates and the accompanying plans I descended on WMC's Melbourne headquarters again and confronted Arvi and Hugh Morgan. The cost of the Research Laboratory appealed to them but they were not enthusiastic about the use to which the money was to be put. They were greatly taken with the alternative which would bear the name of the WMC Conference Centre and as well as serving the School's needs would be a valuable asset to the Kalgoorlie community to which it would be available, but they were a bit aghast at the price tag. They came around to supporting the Conference Centre. A noble

WA School of Mines, August 1959.

action! WMC had never sponsored a project at anything like that expense before and as icing on the cake Arvi agreed to pay the money over in toto immediately. This was a period of high interest rates and combined with the restraint of the builders, Jaxon Construction in submitting their accounts, an additional tax exempt $140,000 accumulated during the design and construction period which paid for several additional items such as a very sophisticated audio-visual system which had not been included in the turn-key contract price.

Simultaneously Odwyn had been active in raising approximately $1,250,000 from the industry in sixty five scholarships most worth $5,000 a year for four years. Between the two of us the response we got from industry was magnificent.

In 1991 Odwyn considered that after fifteen years as Principal of the School he should relinquish the post and leave it to a new man to carry on and hopefully bring some new ideas with him. He had been performing the dual role of Principal of the School and Director of the Brodie-Hall Centre, but despite his enormous energy the task proved to be too great for one person.

The parent body, Curtin University of Technology, came forward with a very helpful proposal. Odwyn would be employed for half of his time as Director of University Development and the other half in his position as Director of the Brodie-Hall Research and Consultancy Centre – an arrangement that has worked satisfactorily as there is a significant synergy between the duties of the two positions.

Relinquishing my position of Chairman of the Board of Management, a personal decision influenced in no small measure by the resignation of Professor Jones, was not accompanied by any great regrets which might otherwise have been the case if the new Principal and the academic staff had been co-operative in the development of the Brodie-Hall Centre.

On Graduation Day 1991 I was approached by Bruno Ballardi, the WASM Student Guild President, to ask if I could "suggest how the students could get new 'on campus' Guild premises". I gave it some thought and at the annual dinner of the School of Mines Graduates Association held that night announced that I would make the Student Guild Building my valedictory project. I gave an off-the-cuff estimate of the cost at $1.5 million which I proposed to raise from graduates of the School and mining supply houses, oil companies, heavy equipment suppliers etc. My estimate of $1.5 million became the actual cost by the simple method of instructing the builder and architect to design and build for that sum.

Melbourne Office (Riverside).
L to R: Gilbert M. Ralph, Hugh M. Morgan, Sir Arvi Parbo on occasion of WMC's 60th Anniversary.

The initial move to raise the money was to put it to four graduates, Graeme Smith, Barry Patterson, Ron Harken and George Botica who, operating as an entrepreneurial group, had done very well out of the 1980's gold boom and had used their money and very considerable mining engineering skills and management capabilities to establish Eltin Mining Ltd a contract mining company, and National Mine Management. Our first meeting was over lunch at my house when after describing the project and its costs I suggested to them that as it would be so much easier to get the money from only one donor they might give consideration to picking up the total sum between them. When Ron Harken turned to George Botica and said, "The old fellow hasn't lost his touch;" I knew I had set my sights too high. Within a week or so I had advice from them that the four families would combine to contribute $250,000. Eltin Mining, a contract mining company, and National Mine Management, which they controlled and had a substantial financial interest in would each donate $250,000. I was halfway there. From then on it was much heavier going. The response from the oil companies was completely negative but going on past records I should have expected that. In particular the graduates were a disappointment. Most of them are in well paid positions and as the contributions could be staged over four years and were tax exempt most could have afforded to make a substantial donation. I think it will be a long time before loyalty to Alma Mater is as strong in Australia as in the United States of America.

I had to admit that I had run my race when the appeal had reached $1,139,572 and some other person would have to pick up the ball and run with it. After many years of fundraising I find that people are either generous or not and that as a generalisation the poorer sections of the community are more generous than the wealthy.

Stage one of the premises was planned to open in May 1994. Stage two will be built when some other enterprising person raises another half a million.

Chapter 18

By 1982 having retired from most of the boards I served on I had ceased to have a regular income other than a couple of lesser boards which paid relatively small fees and as they were quite inadequate to protect my nest egg from the rapid erosion of inflation I started to look around for suitable investments. Suitable in this context meant safe and profitable and it is ironical that I should become financially involved with two grass roots ventures which were highly speculative.

One of these, the ERG Unit Trust, was ultimately developed into the successful business ERG Australia which is still playing an important part in keeping the wolf from my door. The other, Karri Valley a holiday and tourist resort in the South West incorporating a trout farm became a real headache. I had bitten of more than I could chew and found it increasingly difficult to manage. Subsequently they both became the target of an unsuccessful attempt to gain control of them by a pseudo religious sect operating under the name of the Rajneeshees, more commonly known as the Orange people. The activities of ERG, Karri Valley and the Rajneeshees somehow became associated and developed into quite a story.

It began when Dr Mike McCall called into my house one day in 1980 to show Jean and me photographs he had taken of a property in the South West. It was up for sale and he had looked it over. He put it forward as a very attractive piece of land to develop as a holiday resort. The photos were very appealing and I also was very keen to make a personal inspection. Jean admired them but thought that I should not become involved as I was busy enough with the School of Mines and other things and if I decided to go on with it I would not have the time to give it adequate attention. Looking back it was a great pity that I did not heed her advice. We visited the property, I took one look and was hooked.

It was a unique property in that it lay within the State Forest Reserve from which, I was told, it had been excised early in the century by the then Minister for Forests in favour of his daughter. It was subsequently acquired by a man who developed it as a hop farm and in the process constructed a very large dam for irrigation purposes. The dam seen from any position other than a frontal view of the wall that he had built across a valley appeared as a natural feature. Surrounded by Karri forest which reached down to the waters edge, forest and lake merged and created a picture of great beauty, peace and tranquillity.

Mike's idea was to develop a holiday resort to a plan which would be totally sympathetic to these surroundings and would cater only for people of like mind. It would be built up at a pace that we could comfortably fund; a few chalets a year, a camping area for backpackers then, later on, a restaurant and so on. I decided to join him in the project. It wasn't long before we found that we were in for an all weather road system, and a few other odds and ends like water and electricity supplies and sewerage. Supervision of these activities could not be effectively carried on without a manager so one was appointed. The manager had expansive ideas which included purchasing a portion of the farm which had been excluded from the original sale and which had a large hop drying shed on it which, so the manager said, could be inexpensively converted into attractive accommodation either for itinerant holiday makers or large organised groups. We agreed that should be done. The opportunity to establish a commercial trout farm to supply the metropolitan market was too attractive to pass up especially as there was a Scotsman with all the necessary skills in trout farming raring to come out from Scotland to work for us.

Regrettably some of our staff proved unable to resist the temptation to do a little bit on the side for themselves. One chap built a holiday home for himself using our materials and our staff on our payroll, which critically delayed the development of key parts of the resort, and substantially delayed the opening date. Our equipment was hired out locally for personal gain and the trout which were alleged to be dying of oxygen starvation during bursts of hot weather or from any other cause that could be dreamed up were actually being flogged off to the regional hotels. The cost of the hop shed conversion blew out horrifically and one way and another we were accumulating losses at an alarming rate.

It was just as well that during the early stage of the project when we were haemorrhaging money at a great rate, Mike drew my attention to another investment opportunlty to earn equity in a new invention by helping to fund its development. The invention was the intellectual property of three young men. Roger Kent the inventor, Gary Woolterton the technician assisting him, and Jaydon Harman, who was managing the exploitation of their brain child, which was vested in Bridle Falls Unit Trust in which each held a one third equity.

The group operated from very small premises in Subiaco under the name of the Energy Research Group Unit Trust. Previously a shop, the window onto the street had been whitewashed over to give privacy to that room which was used as an office. The room behind was the laboratory work room in which work was currently in progress on a module of the invention, a computer controlled electronic display system with the potential for use in general billboard advertising, or at railway stations and airports for arrival or departure schedules, sports grounds or race tracks etc. The back room held the junk inventors gather around them and, taken at face value, it looked a rather tatty lot. However the invention itself was the essence of simplicity and it was not difficult to dream up even wider avenues for its use.

The prototype worked well and Mike and I were impressed, considered it a good punt and after Mike had negotiated the terms down from $100,000 for a ten percent equity to $100,000 for eighteen per cent interest we each became proud owners in nine per cent of the action for $50,000. That $100,000 put the group in funds and we were invited to be inaugural directors of the Unit Trust and I as Chairman. Shortly afterwards when a good head of steam had been worked up under Jay Harman's very entrepreneurial management and there were half a dozen or more new inventions at every stage of investigation or development Mike confided in me that Jay was Orange. This meant nothing to me. Had he said that he was blue I could have put it down to the cold weather or he could have been green with envy or white with fear, but orange! I could not connect Orange with any emotion or condition.

I don't think Mike knew all that much about it either other than that the Orange people were a strange sect that lived in communes, associated and indulged in free love (a preposterous idea that there is such a thing) wore orange clothes and a medallion thing they called a Mala suspended from a chain worn around the neck in honour of their leader called the "Bhagwan". I have always held very firm views about the rights of people to do their own thing and said in reply to Mike that as far as I was concerned as long as Jay was working within the law, doing his job well and not wearing himself out indulging in an excess of free love he could be any colour he fancied. I should not have taken the matter so lightly. It was not long before the Oranges, officially named Rajneeshee moved to take over both Karri Valley and ERG. I learned more about them as time rolled by.

The sect took its name from its founder, an Indian named Rajneesh Chandra Mohan, who glorified himself with the title of Bhagwan, which I surmise meant leader or top dog. He repudiated the existence of all forms of spiritual authority, God, a supreme being, the likelihood of resurrection, reincarnation or life hereafter although he managed to arrange tax exemption for the sect on religious grounds. His teaching could be fairly summarised as "eat, drink and make love for tomorrow we die". With this philosophy and his encouragement to his followers, known as Sanyasins to "love one another" in a carnal rather than a spiritual manner he was not short of followers.

The delights obtainable from the possession of a lot of money had great appeal for the Bhagwan and rather than amass it by personal endeavours he preferred to extract it from his Sanyasins who collectively contributed on a grand scale. There was an unusually large number of alleged millionaires among them. Before long his sect had become international with branches in nine countries including one in Sydney and one in Perth to which Jay Harman belonged.

Bhagwan founded his sect in Poona India and although there are some very rich people in India they are a very small minority of the population and as the acquisition of great personal wealth appeared to have been his main motivation it is not surprising that he very soon decided to transfer his headquarters to the U.S.A which country – and I mean no disrespect – has a much larger proportion of wealthy cranks than India.

The move was made in 1980. A large tract of land, two hundred and sixty square kilometres, was purchased in Oregon, U.S.A. where he established his headquarters naming his new commune Rajneeshpuram. It was not long before the unseemly behaviour of members of the sect had earned them the opprobrium of the local people and a strong movement developed to eject them. This made headlines in the U.S. but at that time was not of sufficient international interest to become news in Western Australia and when Harman engaged one of the sect, Ma Susheila, (otherwise Mrs Tribe) as a marketing consultant, having assured me that his Orange connection had opened doors to many influential people in the U.S., I was only mildly surprised and raised no objection. With further improvements to the Display System and a sizeable increase to the Unit Trust portfolio of inventions under investigation, the point was reached when it was considered that we should seek a joint venture partner in the Display System as a first step to going public and listing on the Stock Exchange.

A Hong Kong based company specialising in the manufacture of electronic equipment and components controlled by Bill Wyllie was identified with which Harman opened negotiations. Wyllie became interested and visited Perth to look over ERG Unit Trust, its management team and directors, to see the latest prototype and generally to assure himself that the Unit Trust had substance. A manufacturing agreement was signed and preparations made to convert the Unit Trust into a public company. After talking to Wyllie, Hartley Poynton, Stockbrokers, agreed to underwrite an issue of three million fifty cent shares at a premium of twenty cents to raise $2 million. Twenty million fully paid shares were issued to members of the ERG Unit Trust to acquire the Trust Assets. The new company Energy Research Group Australia Ltd was listed on the Stock Exchange in January 1985 and, under the stimulus of Harman's imaginative and purely personal presentation to Eastern States stock brokers of his opinion of the new company's prospects, the share price quickly rose to over $2 a share and Harman, a major shareholder in the company, was being referred to in the daily press as a multi millionaire which on paper no doubt he was.

While this was happening the situation at Karri Valley was rapidly deteriorating without our knowledge of how bad it was. When an irate supplier contacted me directly to complain about the very long delay in the payment of his accounts we became aware that the situation was much worse than the monthly statement of accounts indicated. The manager had not brought to account numerous large invoices by keeping them going to and from between himself and the suppliers claiming that they were incorrect, demanding explanations and more information. An audit revealed the seriousness of the situation. After we had disclosed this to the Bank which was pressing us to reduce our indebtedness we tried to sell the business and take the losses but no one wanted to buy it at an acceptable price. We had not been successful in getting tour operators to put Karri Valley on their itineraries, although as well as the splendour of the lake and its backdrop of noble karris we were able to offer good meals in the Hop Shed restaurant, a trout farm, and there weren't many of them about at that time, and a 'Put and Take' trout fishery where everyone could be guaranteed they would catch four or five trout which they were able to purchase at a very modest price. We then approached Harman who had exhibited a high level of interest in the property. He agreed to come in and acquired a fifty-two per cent interest and was given a seat on the Board and equal voting rights with Mike and me. To satisfy the Bank, Harman deposited a sizeable parcel of ERG shares to increase the company's asset base covering the mortgage.

We replaced the manager with a professional obtained through a consultancy group but he was not able to improve the viability of the resort which at Harman's request we then let to the Rajneeshees who wished to use the Hop Shed as a school. The storm over the Orange people, although raging in Oregon U.S.A., had not yet started to build up in Australia and when Harman suggested that the School should be called the Sir Laurence Brodie-Hall School I was able to refuse in good humour and without any personal rancour. Harman took an immediate interest in his new acquisition and made frequent visits down there. On one of his visits, finding that the manager of the Hop Shed Hostel and Restaurant had given notice, he strongly recommended for the position a woman he knew who had extensive experience in restaurant management.

We engaged her and she performed very well but her appointment caused a regional upset when a Canadian tourist seeking overnight accommodation recognised her as a Rajneeshee from the Mala she was wearing. Having been close to the eye of the storm that was raging between the authorities and the Orange people in Oregon he made it his business to tell the Editor of the Manjimup newspaper about the Oregon experience and what they could expect. He apparently told a good story which the newspaper made the most of. Their enquiries in the U.S. revealed that the Rajneeshpuram had been turned into an armed camp with an arsenal of automatic weapons sufficient to fit out a battalion of marines.

When the peaceful Manjimup community became aware that the Rajneeshees were firmly established at Karri Valley it didn't take them long to put two and two together and recognise that if the Rajneeshees were kicked out of Oregon they would probably establish new headquarters at Karri Valley.

These events gave added momentum to happenings involving ERG. Harman had been away from Australia quite a lot after the company was floated, following up leads in the UK, Europe and the U.S. and it was during a visit he made to the U.S. in March 1985 that the Board became aware of what had been developing between him and the Rajneeshee headquarters.

He left ostensibly to follow up a number of leads in New York. During his absence some Rajneeshees arrived to whom Harman had offered the use of some office space at ERG offices. They went a long way beyond what could be expected when they accepted the courtesy that had been offered them. They entered the building out of normal office hours, night time and weekends, investigated the contents of the filing cabinets and ran up an impressive bill on telephone calls to the U.S. Coincident with this we had disquieting criticism from some Eastern States stock brokers telling us that Harman had misled them about ERG by giving them untrue information on forward contracts and projected revenues. The Hong Kong manufacturers also complained that it was impossible to pin Harman down on projected orders and found him quite unreliable.

When Kent and Woolterton found out what was going on they wasted no time in tracking Harman down, found that he was at Oregon with the Bhagwan, and informed me of this and the behaviour of the visiting Rajneeshees.

At an emergency meeting of the Board called to hear Kent and Woolterton's story, directors decided that Harman would have to go. We got him to a meeting two days later four hours after he got off the plane from the U.S, made our accusation that he was grossly derelict in the performance of his duty, listened to his attempt to refute our statements and then terminated his appointment as Managing Director. Following that action the whole shoddy plot for the Rajneeshees to take over ERG and for the Bhagwan and his entourage to quit the U.S. and take up permanent residence in Karri Valley was exposed.

The Bhagwan had gathered some colourful characters around him in the upper tiers of his management team and I don't just mean Orange colour but unscrupulous ambitious people and foremost among these was Ma Sheila Silverman.

Silverman had been one of the early visitors to the west and had accompanied Harman on one or more of his visits to Karri Valley. I thought she went down to admire the beauty of our South West but later events had indicated that she had more in mind than that.

During that initial visit Harman had shown her off to the ERG Directors at a dinner he gave in her honour at his very comfortable, recently acquired home in Saunders St, Mosman that he had renovated and redecorated for the occasion. It was an interesting event as it showed that the Rajneeshees had brought the caste system with them from India. Ma Sheila and the marketing consultant Ma Sushila were weighted down with heavy gold necklaces and bracelets, with watches glittering with diamonds including one that Harman had bought Sheila in Hong Kong allegedly for $15,000 while the little Sanyasins, young Fremantle girls who had prepared and later served the meal, padded around in bare feet wearing cheap little cotton print frocks and being treated like untouchables. Sheila, exhibiting all the elegance, poise and warmth of a cobra surveying potential victims, allotted places around the dining table to the local Rajneeshee guests in descending order in relation to their potential to be fleeced.

This occurred before any of us were aware of Harman's key role in the Rajneeshee plan to acquire both ERG and Karri Valley and I made a polite speech on behalf of the Board praising Harman for his efforts and success in advancing the Company's business. What a laugh!

After I sacked Harman from his position of Managing Director a small contingent of Rajneeshees which included a brace of lawyers and led by Ma Sheila Silverman arrived in Perth from the USA. They quickly started haunting the ERG offices and used intimidating behaviour with the staff in an endeavour to get access to board minutes, accounts and confidential correspondence. As they had a mole within the staff, changing locks was ineffective and necessitated engaging security guards from Mayne Nickless.

Silverman gave notice of her intentions when she went into print in the weekend press, her statements colourfully illustrated on the front page of the newspaper by a photograph of her, with right hand raised, second finger menacingly extended upward to give point to her threat "I am going to get Sir Laurence". I felt in imminent danger of being raped!

They then directed their efforts to having me dismissed from the Board on the, grounds that I was over the age of seventy-two years, the statutory age for the retirement of a director unless voted in by a seventy-five per cent majority.

In the interval the Rajneesh had endeavoured to raise six million dollars within the ranks of their Sanyasins to get a controlling interest in the company. Fortunately they miscalculated as they depended on having available the shares owned by Jay Harman but which were held in Bridle Falls, a trust equally owned by Harman, Kent and Woolterton and which his two partners would not release. Harman claimed he had an option on Kent and Woolterton's shares but this proved to be invalid as no downpayment had been paid on the options. In the event they got nowhere as they were unable to marshall sufficient money. Their investigation of the ERG minutes however revealed that through a technicality the Board had been illegally appointed during the transition from the ERG Unit Trust to ERG Australia Pty Ltd, the directors having been voted in en bloc without the necessary preceding motion authorising that procedure. There was a very rowdy extraordinary general meeting of shareholders called to elucidate that point, attended by a preponderance of Rajneeshees who had the numbers but not the voting power, at which the

Board was voted out on a show of hands and then re-instated by a poll.

Big trouble for them was brewing in other quarters. After trying to poison Bhagwan and others of his followers, Ma Sheila Silverman and some of her gang skipped the country taking all the liquid assets they could lay their hands on, leaving the Bhagwan with only his fleet of sixty Rolls-Royce cars he had been accumulating.

Silverman was tracked down to Germany from which she was extradited to the U.S. and committed to jail. It only remained then, in respect of my own affairs, to bring Karri Valley to a close.

The resort was sold to the Prindiville Group and Pannel Kerr and Forster engaged as liquidators. The smaller shareholders were paid out, Harman was paid out in full, a precaution against the likelihood that he would make false accusations that he had been used to his financial disadvantage. Mike and I were the main losers. A lesson was learned, not to become financially involved in a business without possessing either the relevant qualifications or background experience.

Karri Valley flourished under the management of the Prindiville group. After being voted back on to the Board of ERG, I relinquished the Chairmanship in favour of Peter Fogarty and shortly afterwards my seat on the board. The involvement with the Rajneeshees followed by the October 1987 share market crash resulted in the shares sliding to a low of seventeen cents.

Careful restructuring by Peter Fogarty and the development of two new products has raised the price of ERG shares (4th February 1994) to $3.60 and the very substantial loss I suffered on the Karri Valley roundabout I recovered on the ERG swings.

ERG now has four hundred and fifty employees and its market capitalisation in February 1994 is $450 million.

Chapter 19

After the tumult and shouting associated with the Rajneeshee affair had died down and the dust had settled, life became dull by comparison and it was rescue from approaching boredom when Mike McCall drew my attention to an opportunity to do something useful for the Australian community and at the same time, hopefully, to lay the foundations of a profitable personal investment.

Besides introducing me to the Energy Research Group and the land at Pemberton as previously described, he drew to my attention a new investment avenue that had been developed by the Federal Government to stimulate new business investment. The concept of Venture Capital Companies had been around for some time and its application in the USA proved to be very successful. As in most things Australia follows America's lead and our Federal Government decided to encourage the formation of a small number of similar businesses in Australia under the name of Management Investment Companies (MICs). These, which would enjoy the incentive of one hundred per cent tax deductibility, would be carefully chosen by Canberra. To keep the players honest there were claw-back provisions for the foregone tax if the investment was disposed of before five years elapsed from the time of investment, the amount of exempted tax subject to claw-back being proportional to the unexpired five year investment period.

I was taken with the idea and believed that as Western Australia is not short of attractive investment opportunities and tax exemption is a powerful incentive I could get financial support for a West Australian based company to raise the minimum capital requirement of $5 million within Western Australia.

While I was in the process of canvassing my friends I found that one of them, Stuart Morgan, had similar ideas and we quickly decided that we could make a stronger case if we joined forces than if we went it alone. As only one licence was to be issued to Western Australia in the first round of what were to be annual approvals of new fund raising, either to top up existing licences or to issue new ones, we approached the State Government to ask if they were interested in taking some equity in the business.

This was in the early years of the Burke Government when WA Inc. was being hatched. Not knowing what sort of a bird would emerge from the egg but realising that State Government participation would put some seal of approval on our application when it was assessed by the Federal Committee set up to administer the scheme, we went ahead. Negotiations were conducted through Mal Bryce, Deputy Premier and Minister for Science and Technology who supported our application when it came before Cabinet and brought the Government in for ten per cent equity. Stuart contacted Newmarket Ltd, an English Venture Capital Company who also came in for ten per cent , and with commitments of a small number of friends we had $2,000,000 (forty per cent) pledged before we went to the public with an offer of shares which Hartley Poynton and Company underwrote. We decided on the name Westintech Corporation. It was not our top priority but when we went to register a name, the combinations of Western and Technology had been thoroughly integrated in a series of names which limited our choice.

We took great care in preparing a powerful application and, to make sure it received the attention it deserved, had a bound copy delivered personally to each of the MIC committee members set up under the Chairmanship of Ralph Ward Ambler. Whether this attention to detail was justified or not we were notified that Westintech's application had been approved.

On setting up business we found that we were not short of people and businesses needing financing. I suppose it was predictable that the very great majority of applications merited only a superficial examination before being rejected. It appeared that every hopeful inventor in Perth had been waiting for just such an opportunity to launch his brain child. Among the remainder there were several that appeared to have the potential to be developed into highly successful businesses. These were mainly established companies whose futures were restricted by lack of capital, but in some cases, when their accounts for past years were investigated, vital financial information was withheld and in the interest of our shareholders these had to be rejected. In other cases, after our investigations had indicated a bright future in which we were justified in investing, the proprietors had gained sufficient confidence to seek the necessary money through the stock exchange and float, usually for second board listing.

However we did support several projects and in the ensuing four years raised a further $6,770,000 through calls. Some failed, others benefited from the capital injection they received but without proving a bonanza investment for Westintech.

In June 1987, having reached the mature age of seventy-seven, I decided I should resign from the Board and let someone else, preferably a few decades younger take over. I wasn't influenced in this decision by the lack of success of the company. Venture Capital Companies are highly speculative and rely on the occasional major success to compensate for those that fail or just bumble along and people know that when they subscribe. If they wanted security they would put their money into Government Bonds.

In March 1987 Stuart Morgan and his very able lieutenant Kyle Abbott had successfully signed a contract on behalf of Westintech with the Aeronautical Division of Rolls-Royce to manufacture turbine blades for aeronautical jet engines. This was a great triumph for them because there was very strong competition from companies in other states for the business. The new company named Turbine Components Australia Ltd was owned fifty per cent by Rolls Royce and fifty per cent by Westintech. The new company built premises at Cannington to house three furnaces which would operate round the clock with a contract to sell the output for five years to Rolls-Royce. Within two years production had been tripled with the construction of six additional furnaces.

This increase had been gained through the influence of Ansett Worldwide Aviation Services, an aeroplane leasing company in which Ansett Australia had a fifty per cent interest. Ansett had secured a significant interest in Westintech through the purchase of Circuit Technology and when Ansett in turn made an offer for Westintech in August 1990, although it was a miserable offer, $1.20 a share, the directors decided to accept. Westintech shareholders made a small profit. The $1 shares because of the tax exemption had a real cost of a little over fifty cents. The real beneficiary was the State which then had a brand new business with an internationally strong partner in a secure and rapidly growing technology.

T. C. A. appear to be settling down quite nicely. Accounts of June 30 1993 show sales volume of $28.1 million and an after tax profit of $5.4 million.

Subsequently Stuart Morgan went on to develop a new business producing super alloys used in the manufacture of aeronautical jet turbine blades. This new venture situated on land adjoining Turbine Components Australia Ltd called West Australian Super Alloys is fifty per cent owned by Western Aero Space, a West Australian company, twenty five per cent by Pratt and Whitney the world's largest manufacturer of jet aero engines, and Wyman Gordon, an American company specialising in the production of super alloys. The plant was commissioned in September 1993. Its initial production will be three million seven hundred and fifty thousand pounds of alloy per year which I expect will rapidly expand.

☆ ☆ ☆

In 1962, during my Kalgoorlie period, I was asked if I would be interested in an appointment to the State Committee of the Commonwealth Scientific and Industrial Research Organisation (CSIRO). It seemed to be something that I could fit in with my normal work commitments and which held promise of providing useful contacts within CSIRO and I accepted. I found that I could cope with the meetings which tended to be irregular, informal and somewhat unproductive.

It is probable that during the next ten years of service on the committee I thought of ways in which it could be made more productive and when, in 1972, I was offered the Chairmanship I accepted. However my acceptance was with the proviso that it should become more formal and meet bi-monthly on fixed predetermined dates, that its objective should be to investigate the research needs of key areas of Western Australian industry, that a report would be prepared by a subcommittee of experts on the subject being investigated and submitted to the Chairman of CSIRO with the request that it become an agenda item for consideration by the CSIRO Advisory Council at the meeting next following receipt of the report.

I personally canvassed for new members for each committee among whom were Bruce Kirkwood, Chairman of the SEC, Noel Fitzpatrick, Director of the Department of Agriculture, Bob Hillman, Director of Public Works, Professor Street, Vice Chancellor of the University of Western Australian and many other community leaders. It was a matter of personal satisfaction that everyone I approached agreed to join the committee.

Members were pleased that in future the work of the committee was to have a focus and agreed to support the new programme. The general format was an afternoon meeting at two monthly intervals which ran for about two hours at which a speaker, usually a senior officer from CSIRO, presented a paper followed by discussion and/or the selection of the next research project.

Once the subject was decided, membership of the project committee was drawn from the scientific and industrial community at large. Unless there was a member or members of the State Committee with expertise relevant to the project, State Committee representation on the Project Committee was confined to myself and John Brophy, the CSIRO regional administrator, who was also the State Committee Secretary. I reserved the position of Chairman of each Project Committee to myself so that I could ensure that its direction and progress was maintained. This system was very strongly supported by government, industry and academia. The quality of membership of the project committee which invariably had one or more government departmental directors represented on it as well as University Professors from relevant faculties and senior executives from industry, gave weight and credibility to the reports which earned a high level of respect from the CSIRO executive and were accorded detailed consideration by the Advisory Council.

The first project was an investigation into the Multiple Land Use of the South West of Western Australia. This was a very topical subject. The environmental movement was gathering strength and becoming vocal about the effect of bauxite mining in the Darling Ranges. The timber industry was concerned that bauxite mining would encroach into their domain which was under serious threat in the long term from jarrah dieback and CSIRO itself had reservations on the effect of bauxite mining on water quality. That initial project gave the revitalised State Committee a good start and resulted in CSIRO posting some scientists and technicians to WA to monitor water salinity in mined areas and areas adjacent to bauxite mining.

During the ensuing nine years the State Committee produced a report annually on a variety of topics ranging from "The Effect of Rye Grass Toxicity in Sheep", "The Economic Potential of Fruit and Vegetable Production in the Kimberley Region", "The Western Rock Lobster Industry", to "An Oceanographic study off the North West Coast of Western Australia". This latter was of great importance to the oil and gas industry in the design of off-shore engineering structures which would be subjected to the stresses of strong ocean currents flowing in different directions at the surface and at depth. It was an excellent report and I thought for a while it would be funded but it came back on the grapevine that it was the reference to "Western Australia" in the title that caused it not to be supported.

The project which generated the most interest was an investigation into the supplementation of the water supply for the metropolitan area of WA with icebergs from the Antarctic.

When Saudi Arabia began to enjoy the full benefit of the massive rises in the crude oil price imposed by the OPEC in the

1970's it turned its thoughts to improving both the quantity and quality of its domestic water supply by supplementing it with Antarctic ice and entered into an arrangement with a French engineering firm, to establish a company called Icebergs International. The Saudis provided the finance and the French the engineering skill under the direction of a Frenchman named Georges Mougin. Their investigations indicated that the best procedure to achieve this was to capture icebergs of appropriate size at the approximate longitude of South America, take them to Reunion Island in the southern waters of the Indian Ocean, where they would be converted to a slurry by mining the ice at the exposed surface and transferring it to bulk tankers for transport over the equator for delivery to a Gulf port. An opportunity was being sought to test the first stage of this operation at a location which would not require large capital sums to conduct a satisfactory trial and Perth was selected.

The initial discussions on this project were held between Georges Mougin, representing the principals of Icebergs International, and the West Australian Committee of CSIRO. The offer appeared to be most attractive. Icebergs International undertook to deliver without cost to the West Australian government an iceberg of a size sufficient to supply a regular daily volume of water in ice-slurry form for an estimated period of four years.

The prevailing westerly winds combined with the ocean currents at Antarctic latitudes give rise to a propulsive force trending north easterly estimated at about 3 knots. Their plan was to capture an iceberg of appropriate size at the longitude of Cape Town and maximise the benefit of these natural forces during transportation on the first leg of its journey. It was proposed to accelerate and control the direction of this motion by mounting sails on the iceberg. A team of men would be landed on the 'berg by helicopter with all the gear for a protracted stay. Their job would be to socket some metal masts into the 'berg upon which sails would be rigged. That team would remain on the 'berg for the rest of the journey to exercise some control over its direction.

When these natural energy sources ceased to impart sufficient velocity in the desired direction, ocean going tugs would be attached to tow the iceberg to its final destination.

Icebergs International offered to deliver an iceberg to the Rottnest Trench where they would anchor it and protect it from excessive erosion from warm water currents by placing protective shrouding around it. The WA Government would then be required to take over. It was considered that the iceberg would be mined at the surface by the equivalent of continuous coal mining equipment which would produce a very coarse slurry to be pumped into a tanker for transfer to the Kwinana Power Station where it would enter the cooling system and generate a further stage of flash distillation. Ultimately, when its latent energy had been released, the water would be fed into the Metropolitan and Country Water Supply System with the obvious advantage of diluting the high concentration of dissolved salts arising from the inclusion of copious amounts of mineralised underground water.

When the Water Supply Authority looked into this, although they recognised the perceived advantages, the economics were not sufficiently favourable to justify the considerable expense that would be involved for only a short term (four year) benefit. I guess the further development of the proposal was not dependant on the proposed West Australian test but as nothing has yet developed internationally it can perhaps be concluded that the Saudis came to the same conclusion as the WA Government.

I was sorry that it failed to go ahead. Water is becoming increasingly a commodity in short supply. With the world population increasing at present rates normal sources of supply will rapidly prove to be inadequate. Other means of supplementation require a significant energy input and should be avoided for economic and environmental reasons. With over eighty per cent of the world's fresh water residing in Antarctica I have a feeling that the continual flow of icebergs that enter the sea and dissolve to no good purpose will ultimately be used along the lines proposed by Icebergs International.

In the late seventies the Federal Government appointed Prof Arthur Birch from the National University, Canberra, to head an enquiry into CSIRO. The State Committee prepared a thoughtful submission which recommended strengthening the Advisory mechanisms of the national organisation. We (the State Committee) met with Arthur Birch over a meal at University House by the end of which his committee appeared favourably impressed with our presentation. The Professor's ultimate report included the majority of our recommendations.

In 1982 the Federal Government in its wisdom decided that any one over the age of 65 was too senile to serve on a government committee and to put it bluntly I was kicked out of my position as Chairman of the State Committee of the CSIRO and the Advisory Council. Sir Ian McLernnon fell under the same edict. I couldn't understand the wisdom of such a direction. Maybe it was just my personal vanity. It was mollified by the establishment by CSIRO of an Annual Lecture titled the Brodie-Hall Lecture which carries on the intention behind the work of the WA State

2 SEP 1981

CSIRO

Limestone Avenue, Canberra, ACT

The Chairman

Dr J P Wild, CBE, FTS, FAA, FRS

PO Box 225, Dickson, ACT 2602
Telephone (062) 48 4621

25 August 1981

Dear Brodie,

I learn with considerable regret that your long and distinguished term as Chairman of the WA State Committee as well as your membership of the Advisory Council comes to an end at this time. I did not want the opportunity to pass without expressing my deep personal gratitude for your untiring efforts, and the stimulation of others, on behalf of the Organization. I can say privately – but not publicly! – that during your term of office the WA State Committee has been quite the

Commonwealth Scientific and Industrial Research Organization, Australia most vigorous and effective. You have become so much the personification of CSIRO in the West that it's going to be difficult to adjust to the new situation.

With warmest best wishes

Yours sincerely

Paul

Committee, to draw to the attention of the general public the splendid work carried out by their National Scientific and Research Organisation.

For 10 years as Chairman of the W.A. State Committee, I was an ex officio member of the National Advisory Council of CSIRO which met quarterly in the Eastern States. This provided an opportunity for me to form a close association with the Chairman and members of the CSIRO executive. The friendships formed continue today.

More recently State Committees around Australia have been terminated. The two Divisional Chiefs that were appointed to WA, largely at the instigation of the State Committee, have since been moved back to Eastern Australia! With the extinction of the State Committee a conduit between the State and CSIRO was broken. My association with the Committee and the CSIRO generally led me to believe in the need for a close relationship between the state and our National Research Organisation.

WA State Committee Reports prepared under my Chairmanship:

1. Proposals for environmental studies into some aspects of natural resource utilisation in South West Australia.
2. Proposal for marine environmental studies off the West Australian coast.
3. Systems analysis studies in resource management for Western Australia.
4. Long term trends in Agriculture in Western Australia.
5. Understanding the near surface environment of the Kalgoorlie Region to aid mineral exploration and water management. An opportunity for CSIRO Research.
6. Research problems in Western Australian sheep fertility.
7. Observations in cloud seeding in Western Australia.
8. Review of forests and forest products problems.
9. Studies in tropical ecology. A multi disciplinaryapproach to arbovirus in Western Australia.
10. Applied Physical oceanography. A preliminary study into Western Australia's needs and future research programmes.
11. New land use options for the South West of Western Australia.
12. Fresh fruit and vegetable production for North West Australia.
13. A proposal for a national water supply technology demonstration centre in Western Australia.
14. A submission to the Committee of enquiry into CSIRO.

Roy Woodall measuring himself against a blackbutt during the clean-up.

Rolling logs uphill.

CONCLUSION

The events in this final chapter hardly qualify as memoirs, they are too recent. I have included them to pull together some loose ends and give the story a rounded ending.

At this time I have relinquished all but two of my community activities, International Chairman of the John Curtin Centre Appeal and Chairman of the West Australian Foundation for Contemporary Science and Technology otherwise know as the Scitech Discovery Centre and am on the verge of dropping them off also. This is in keeping with my long held opinion that old people should quit their residual responsibilities before their faculties begin to fail, a condition which it is difficult for the principal to recognise in himself and about which, like b.o., even your best friend won't tell you. The trigger for the soul searching that convinced me that I had reached that stage of life was an essay by Northcote Parkinson contained in his entertaining book of satirical essays Parkinson's Law.

Having lost the fine edge of my hearing when operating pneumatic drills underground or working in close proximity to stamp batteries I am reduced to wearing a hearing aid. Parkinson describes my condition admirably when he had one of his characters comment "The acoustics in this hall are simply terrible. You would have thought these scientific chaps could do something about it. For half the time I can't hear what is being said, can you?" In a Chairman's role, I am in a somewhat better position than that but I find that it doesn't endear me to Board or Committee members to have to exhort them to "please stop mumbling and speak up".

In anticipation of the day when I would no longer have the activities that have kept me occupied and happy for so many years I searched around for a suitable substitute. Shortly after Jean and I were married I bought a piece of land in the hills a few kilometres east of Harvey. Topographically it was very attractive, the ground was gently undulating and sufficient coppices of trees had been left standing to lend character to the landscape which was further enhanced by a permanent brook. At ninety-eight acres it was just the right size. When I bought it it was very untidy, with stumps and huge logs dotting the paddocks which took me the next six or seven years to clean up. I bought a large chain saw and enjoyed the challenge of docking the big logs into manageable pieces for burning and then manoeuvring them into position when fire had reduced them to the size at which I could get them around the stumps to burn them out. Most of the trees on the property had been felled down-hill from the stump. I needed all the help I could get as year by year my muscles were shrinking and when I was about seventy-five I realised that I should find an occupation a bit easier then inching logs up hill with a crowbar. For the first three years we lived in a caravan, but later decided that the money saved did not balance out the discomfort suffered and we built a cottage which also enabled us to develop some social activities and have friends visit us over the weekend. Many of my visitors "sang for their supper" by helping me with the cleaning up. Neil Tuckwell and John Brophy had reason to remember their visits, their aching muscles being a painful reminder.

I was taken with the thought of taking up woodwork, a very optimistic idea as years before I had tried my hand at it and failed miserably. The occasion was during the early days of my marriage to Dorothy when we were living at Norseman. Dorothy had intimated that she would like a dining table and chairs and, as we had no money, I decided I would make them. It seemed sensible to start by making a chair as I would not then be committing as much money in materials as would be required if I started on the table. I had never demonstrated great precision at the manual training classes at school where I had the benefit of a good array of tools and a carpenters bench and it was not surprising that my first chair, made with the minimum of tools on the kitchen table, exhibited joints which gave the strong impression that they had been made not with a hand saw, hammer and chisel but with a bushman's saw and blunt axe. Unwilling to admit defeat or, if I continued, to suffer the ridicule of friends at my primitive efforts, I decided to conceal the bad workmanship with a layer of masonite (fibre board). This achieved the desired visual result but drastically altered the chair's centre of gravity. It looked good when backed against a wall but in any other situation gravity took over and it would fall over backwards. There are very few things as useless as a chair that won't stand up.

The author in his workroom and (below) the workroom.

Hand carved "Mallee" root. The snake rampant.

Hall table made from a "Jarrah" burl

Turned "Blackboy" bowl

Carved "Red Morrell" bowl

I always try to balance realism against optimism and on this occasion concluded that I lacked the necessary skills and abandoned the project and with it all thoughts of ever becoming a carpenter.

However I have always harboured a secret belief that I had some latent artistic talent just waiting for an opportunity to emerge. Wood carving seemed to be a possibility. It is a craft requiring only basic tools. I had seen Africans sitting at the roadside, using little more than sharp knives, carving splendid replicas of elephants, buffalo, rhinoceroses etc. and although I did not aspire to reach their level of craftsmanship, using the example as inspiration, I decided to try my hand.

Sculpting is a relaxing occupation. Wood is a pleasant material to handle, it has a friendly comfortable feel and during roughing out it is possible to let the mind slip into neutral or to follow whatever flight of fancy stirs the imagination. Later on I bought a lathe but did not get the same satisfaction as from free form carving. Of course the quality of the wood has a bearing on the end product and I am lucky that I have friends around the state who regularly send me good specimens on which to try my skill.

One of my earliest pieces was a snake, rampant, rising out of a mallee root. I entered it in the Harvey Agricultural Show and won the second prize of $2.00! A hall table designed in such a way that the joints did not show was quite successful and now adorns my son Charlie's house in Esperance. My present project is carving a blackboy stump into a lotus flower. Jean, who has always spoiled me with loving kindness, decided that I should have better premises than the tin shed in which I did my woodwork and exercising her initiative had the local builder erect a very fine workshop which we now dignify with the title of 'Brodie's Studio'!

Jean has her own workroom at the farm in the basement of the house where she intends to develop her own artistic skills working with silver. Until recently she has got her enjoyment from the farm by running a small herd of cattle, twenty-five breeding cows and a bull. The progeny were sold as yearlings except for one that we had slaughtered and butchered for home consumption. As Jean never used herbicides, pesticides or steroids in her animal husbandry she was satisfied that she was feeding me on the healthiest meat available.

We have each travelled extensively in the past both before and since we married and neither of us is consumed by a burning desire to do much more. Recently we made two trips to Britain and the United States to visit my sole remaining English relation, a niece who lives in South Wales, and to continue on through the United States to catch up with a number of friends I made while I was active in the mining industry and some friends of Jean's resulting from the several years when she was the delegate for the Australian Institute of Landscape Architects attending International conferences.

We now propose to stay home and gather some moss.

GLOSSARY OF MINING AND PETROLEUM TERMINOLOGY

Anomaly: A deviation from the norm in rock patterns, usually discovered by geophysical methods, which suggests the possibility of a mineral deposit.

Anticline: An arch or fold in the layers of rock shaped like the crest of a wave.

Back: The ceiling of any underground excavation.

Ball Mill: A piece of milling equipment used to grind ore into small particles, which uses steel balls as the grinding medium.

Base Metal: A commercial metal such as copper, lead or zinc. The term was coined to describe a metal 'inferior' to precious metals such as gold and silver.

Bauxite: An aluminium oxide, the principal ore of alumina.

Biological Leaching: A process for recovering metals from low-grade ores by dissolving them in solution, the dissolutions being aided by bacterial action.

Bullion: Gold or precious metal in bars or ingot form.

Cage: An elevator-type conveyance which moves men and materials up and down a mine shaft.

Carbon-in-pulp (CIP) Carbon -in-leach (CIL): Methods of gold extraction from cyanide solutions using carbon as a precipitating agent.

Chute: A vertical opening in mine through which ore is moved to waiting conveyances, such as mine cars or conveyor belts.

Collar: The term applied to the timbering or concrete around the mouth of a shaft, also used to describe the top of a drill hole.

Concentrate: To treat ore so that the result – 'concentrates' – will contain less waste and a higher amount of the valuable mineral. In many mining operations, ore is concentrated in a concentrator or mill on the surface, then shipped to a smelter, whose product is then sent on to a refinery for the recovery of the mineral content by eliminating impurities.

Core: A cylindrical piece of rock that is extracted from the earth by a diamond drill, and the cylindrical core removed to the surface for examination and analysis.

Country rock: The rock surrounding the orebody.

Crosscut: A lateral or horizontal opening made underground, usually leading from a vertical shaft toward an ore body or vein, and cutting across the direction of the rock or the ore body. (See Drift.)

Cut and Fill: A method of mining where ore is removed and the resulting excavation filled with waste to support the stope before the next section is mined.

Decline: A sloping tunnel providing road access from the surface to underground mine operations.

Development: (mining) Preparation of the deposit for commercial production including installation of plant and machinery and the construction of all related facilities.

Diamond Drill: A type of rotary rock drill which cuts a core of rock in long cylindrical sections, an inch or more in diameter.

Dilution: An estimate of the amount of waste or low grade mineralized rock which will be mined as part of normal mining practices in extracting an orebody.

Dip: The angle at which a vein structure or rock bed is inclined from the horizontal measured at right angles to the strike.

Drift (or Drive): A horizontal underground tunnel in such a direction that it follows or 'drifts' with the ore or an ore vein or favourable rock. (See Crosscut.)

Drill: A machine for making holes in rock. Common types include churn, diamond and rock rotary drills. They may be powered by compressed air, electricity, gas or diesel motor.

Electrolytic Refining: The process of refining metals by conducting an electric current through a solution with the metal being deposited on the cathode.

Exploration: The prospecting, diamond drilling and other work involved in searching for ore bodies.

Fault: A surface or zone of rock fracture along which there has been displacement.

Feasibility Study: A definitive engineering estimate or calculation of all costs, revenues, equipment needs and production likely to be achieved when a mine is developed. The study is used to define the economic viability of a project and to support the search for project financing.

Flotation: A recovery process by which valuable minerals are separated from waste minerals to produce a concentrate.

Footwall: The wall or rock on the underside of a vein or ore structure.

Free Milling: Gold ore containing free gold which can be extracted without the need of roasting to remove sulphides.

Gangue: The worthless minerals associated with valuable minerals in an ore deposit.

Grade: The amount or weight of metal or mineral present in the host rock. Often expressed as percent, ounces per ton, or parts per million.

Grinding: Reducing mineralized rock to the consistency of sand by crushing and abrading in a rotating steel grinding mill.

Hanging Wall: The wall or rock on the upper or top side of a vein or ore deposit.

Head Grade: The average grade of ore fed into a mill.

Heap Leaching: Extraction of gold from heaps of mined rock by irrigating with solutions such as cyanide.

Hoist: A machine which raises and lowers the cage and skips in a mine shaft.

Hydro-metallurgy: Separation of the metal in aqueous solution from the rest of the ore, followed by precipitation in metallic form.

Lagging: Planks or timbers usually placed under the roof of a stope or drive to prevent rock from falling.

Leaching: The removal of minerals from ore by their dissolution in water and subsequent percolation. May be a natural process on in situ deposits or a chemical process used in milling.

Level: The horizontal development of drives and crosscuts from a mine shaft.

Lode: A mineral deposit in solid rock.

Matte: The output of a smelter being metal containing some impurities requiring refining to obtain the pure metal.

Metallurgy: The science of extracting metals from ores and preparing them for use.

Mill: A plant, usually at the mine site, which concentrates ore or treats it so that minerals are separated and prepared for ultimate recovery in pure form.

Mineral: A naturally occurring element or chemical compound.

Mineralization: A concentration of valuable minerals.

Mineralized Zone: Any mass of host rock in which minerals of potential commercial value occur.

Mullock: A low grade material broken during mine development.

Open Pit: A surface working, open to daylight, such as a quarry. Also referred to as open cut or open cast mine.

Option: A right to have the first chance to buy or refuse to buy a property or an asset of some kind. In mining, it often refers to a contract that covers a mining claim or groups of claims.

Ore: Rock from which a mineral, or minerals, of economic value can be extracted.

Ore Reserves: The tonnage of ore of a certain grade which is estimated for a mining property or certain sections of it.

Outcrop: Solid rock or mineral that is exposed on the earth's surface and can be examined visually.

Over-burden: Waste rock overlying an ore body.

Oxidised ore: Ore altered by surface waters and air.

Pilot Plant: Equipment set up on a small scale to duplicate a practical production plant. Its purpose is to test a process prior to the commitment to build a full-scale plant.

Porphyry: An igneous rock in which a fine-grained matrix contains scattered coarse-grained crystals (or phenocrysts).

Pyrite: An iron sulphide mineral.

Pyrrhotite: An iron sulphide mineral containing less sulphur than pyrite.

Raise: A vertical or inclined underground tunnel that has been excavated from the bottom upward.

Recovery: The amount of mineral that is separated and recovered in a mill, expressed as a percentage of that calculated to be in the original ore.

Reef: A gold-bearing vein.

Refining: The final purification process of a metal or mineral. (See Electrolytic.)

Refractory: Ore that resists normal chemical processing and must be roasted to effect reovery of valuable minerals.

Resource: An occurrence of mineralized rock which is inadequately proven or insufficiently mineralized to be classified as an ore reserve.

Roasting: Treating ore, or a partly processed ore product, by a combination of heat and air, to further remove impurities such as moisture or sulphur.

Rock: Any naturally formed combination of minerals constituting an appreciable part of the earth's crust.

Shaft: An opening cut downward from the surface for transporting personnel, equipment, supplies, ore and waste. It is also used for ventilation and as an auxiliary exit. It is equipped with a surface hoist system which lowers and raises a cage in the shaft as well as 'skips' or containers for bringing up ore or waste. A shaft generally has more than one compartment.

Shear: A fracture formed when the portion on one side of the fracture has slid past the portion on the other side.

Shoot: The part of a vein bearing valuable ore.

Skip: A self-dumping type of bucket used in a shaft for hoisting ore or rock.

Stope: An opening underground in which ore or waste is blasted and broken.

Stoping: The extraction of ore in an underground mine by excavating openings in a series of levels or steps.

Sulphides: A compound of sulphur with more than one element.

Sump: An excavation for the purpose of catching or storing water. The bottom of a shaft is commonly used for this purpose.

Tailings: Waste material from a mineral-processing mill.

Vein: An opening, fissure or crack in rock containing mineralized material.

Wall Rock: The rock forming the walls of an underground opening.

Waste: Material that is too low in grade to be of economic value.

Winze: An underground opening similar to a shaft but not starting at the surface.

Write-down or Write-off: An amount recognized to be a loss in value of an asset below its recorded cost, or expenditures relating to an event or transaction which has already occurred yet offers no corresponding future benefit. A cost included in the calculation of the company's earnings for the accounting period.